Silica-Based Organic–Inorganic Hybrid Nanomaterials

Synthesis, Functionalization and Applications in the Field of Catalysis

Sustainable Chemistry Series

ISSN: 2514-3042

Series Editor: Nicholas Gathergood *(University of Lincoln, UK)*

The concept of Green Chemistry was first introduced in 1998 with the publication of Anastas and Warner's "12 Principles of Green Chemistry". Today, these principles are becoming adopted as general practice in the chemical industries in order to reduce or eliminate the use and generation of hazardous materials, reduce waste, and make use of sustainable resources. New, safer materials and products are being released all the time. Alternative technologies are being developed to improve the efficiency of the chemical industry, while reducing its environmental impact. Sustainable resources are being investigated to replace our reliance on fossil fuels – not only as source of energy but also a source of chemicals — be they feedstock, bulk, or fine. Consideration is now given to the whole life cycle of a product or chemical — from design to disposal. And, as more of the Earth's resources become scarce so new alternatives must be found.

As the world works towards meeting the needs of the present generation without compromising the needs of the future, this series presents comprehensive books from leaders in the field of green and sustainable chemistry. The volumes will offer an excellent source of information for professional researchers in academia and industry, and postgraduate students across the multiple disciplines involved.

Published

Vol. 4 *Silica-Based Organic–Inorganic Hybrid Nanomaterials: Synthesis, Functionalization and Applications in the Field of Catalysis*
by Rakesh Kumar Sharma

Vol. 3 *Functional Materials from Lignin: Methods and Advances*
edited by Xian Jun Loh, Dan Kai and Zibiao Li

Vol. 2 *Furfural: An Entry Point of Lignocellulose in Biorefineries to Produce Renewable Chemicals, Polymers, and Biofuels*
edited by Manuel López Granados and David Martín Alonso

Vol. 1 *Sorption Enhanced Reaction Processes*
by Alírio Egídio Rodrigues, Luís Miguel Madeira,
Yi-Jiang Wu and Rui Faria

Sustainable
Chemistry
Series

Volume 4

Silica-Based Organic–Inorganic Hybrid Nanomaterials

Synthesis, Functionalization and Applications in the Field of Catalysis

Series Editor

Nicholas Gathergood
University of Lincoln, UK

Rakesh Kumar Sharma
University of Delhi, India

World Scientific

JERSEY · LONDON · SINGAPORE · BEIJING · SHANGHAI · HONG KONG · TAIPEI · CHENNAI · TOKYO

Published by

World Scientific Publishing Europe Ltd.

57 Shelton Street, Covent Garden, London WC2H 9HE

Head office: 5 Toh Tuck Link, Singapore 596224

USA office: 27 Warren Street, Suite 401-402, Hackensack, NJ 07601

Library of Congress Cataloging-in-Publication Data
Names: Sharma, Rakesh Kumar, author.
Title: Silica-based organic–inorganic hybrid nanomaterials : synthesis, functionalization and
    applications in the field of catalysis / Rakesh Kumar Sharma (University of Delhi, India).
Description: pages cm | Sustainable chemistry series, 25143042 ; 4
Identifiers: LCCN 2019943869 | ISBN 9781786347466 (hardcover)
LC record available at https://lccn.loc.gov/2019943869

British Library Cataloguing-in-Publication Data
A catalogue record for this book is available from the British Library.

For any available supplementary material, please visit
https://www.worldscientific.com/worldscibooks/10.1142/Q0221#t=suppl

Desk Editors: Herbert Moses/Jennifer Brough/Shi Ying Koe

Typeset by Stallion Press
Email: enquiries@stallionpress.com

Foreword

This book presents the contemporary topics to academia and industry scientists. Entitled *Silica-Based Organic–Inorganic Hybrid Nanomaterials: Synthesis, Functionalization and Applications in the Field of Catalysis*, this book features several aspects regarding strategies used for various nanomaterials and their applications in catalysis. There are seven outstanding chapters featuring exhaustive coverage of nanomaterial allied sciences. It is great to find that the authors successfully attempted to establish a connect between Green Chemistry and Nanomaterials. The elegance of the book is backed by the development of silica-based nanomaterials spiced with the magnetic properties and their profound applications in catalysis. Despite the ready availability of many journals and access to various databases, search engines and books, it is still difficult for chemists to routinely and quickly glean informed inputs for selecting the right strategy to arrive at a desired nanomaterial output. This book, managed by practicing chemists with a wide range of chemistry backgrounds, one is an experienced and highly reputed chemist from university involved in research and teaching and few others are senior research scholars, captures outstandingly notable aspects where synthetic strategies for nanomaterials are sought to be more practical. Valuable information about existing and upcoming strategies for diverse ways of preparing the highly defined nanomaterials is also included in the book. The extent of information and trends in nanomaterial sciences captured in this book make it an essential resource for scientists working at the frontiers of nanomaterial research both in industry and in academia. All the salient features are captured with considerable rigor and duly factor in the context of each example and the choice of strategies employed in a particular isolation. All the strategies provide the relevant background and include relevant references. The book can be regarded as a benchmark for nanomaterial research. Taken as a whole, this contribution has the potential to offer an extremely useful guide to arrive at more practical and relevant strategies for nanomaterial preparations and their applications in the area of catalysis. The contributors of the book, Prof. R. K. Sharma and various research scholars, are from the Department of Chemistry, the University of Delhi, Delhi. It is to be hoped that this endeavor will enthuse scientists in nanosciences working at industry and academia to revisit and devise strategies for applicable nanomaterial preparations that are more practical and eco-friendly.

Rakeshwar Bandichhor, PhD, FRSC, CChem,
Director, API-R&D, Dr. Reddy's Laboratories Limited,
Hyderabad, India

Preface

In the recent years, significant advancements have been made in the synthesis of well-defined materials that can be employed as heterogeneous catalysts for various types of organic transformations because they enable environmentally friendly and benign catalytic processes. In this regard, the field of nanocatalysis is undergoing many exciting developments and the design of silica-based organic–inorganic hybrid nanocatalysts is a key focus of the researchers working in this area.

This book aims to present a succinct overview of the recent research progress directed toward the fabrication of silica-based organic–inorganic hybrid catalytic systems encompassing the key advantages of silica nanoparticles (SNPs) and silica-coated magnetic nanoparticles in an integrated manner. Through this book, we present the judicious design of such types of highly active and selective nanostructured catalysts *via* the simple manipulation of the interaction between the catalytically active nanoparticle species and their support for the very first time.

The book showcases a detailed description of the development of silica-based organic–inorganic hybrid nanomaterials that have captivated the attention of several researchers worldwide owing to their outstanding catalytic properties, such as enhanced thermal stability, increased selectivity, ease of isolation and greater reusability. It also enlightens the readers about the different types of approaches involved in the synthesis, functionalization and characterization of such types of hybrid nanomaterials. Remarkably, this book also provides an enormous amount of knowledge to the readers about the fusion of nanotechnology with Green Chemistry that strives to meet the scientific challenges of protecting human health and the environment.

Hallmark Features

- A *unique combination of silica and silica-coated magnetic nanoparticles* is presented in a holistic manner for catalyzing diverse industrially significant reactions.
- A *discussion on all synthetic approaches* enlisted in the literature (including some recent methodologies such as flow- and microwave-assisted synthesis) that enable large-scale synthesis is provided, thus proving to be useful not only for the academicians but also for the industrialists. It also plays a crucial role in tuning the morphology and consequently tailoring the properties of the nanomaterials.
- A *deep insight is provided into the procedures adopted for the surface modification* of silica and silica-encapsulated magnetic nanomaterials (throwing light on some excellent functionalization methodologies such as *in situ* and post-synthetic functionalizations) that smartly tune the surface properties of these nanomaterials, rendering them suitable for catalytic applications.
- A *detailed description is included on the physicochemical tools* utilized for characterizing the nanomaterials that assist in evaluating the morphological, structural and characteristics of the developed catalytic materials.
- *Examples of numerous industrially significant reactions* that are catalyzed efficiently by these nanomaterials are presented.
- A *critical comparison* of silica-based nanocatalysts with other supported and homogeneous catalysts is presented, taking into account not only their advantages but also some drawbacks that would broaden the outlook and enhance the critical thinking of the readers by helping them understand the catalytic processes.
- A *special emphasis is provided on the future perspectives*, i.e. areas that can be significantly improved, such as rational design of multifunctional nanocatalysts, robust and recyclable catalysts

that can effectively perform multiple catalytic reactions with high atom efficiencies, yields, chemoselectivities and enantioselectivities within a one-pot system.

It is anticipated that this book shall be an excellent contribution to the scientific community worldwide as a single book would combine almost all the crucial aspects associated with the precise design and engineering of such type of hybrid catalytic systems. Also, it is expected that the book would eventually promote sustainable organic synthesis by opening up new opportunities for the large-scale synthesis of several industrially significant products, thereby benefiting a broad range of both academic as well as industrial readers.

I acknowledge the contributions of the co-authors, Dr. Yukti Monga, Dr. Shivani Sharma, Dr. Manavi Yadav, Dr. Sriparna Dutta, Ms. Rashmi Gaur, Ms. Radhika Gupta and Ms. Gunjan Arora from Green Chemistry Network Centre, Department of Chemistry, University of Delhi, Delhi-110007, India.

R. K. Sharma

About Professor R. K. Sharma

R. K. Sharma is the Coordinator of Green Chemistry Network Centre, established under the recommendation of World Leaders of Green Chemistry headed by Prof. Paul Anastas. He is a fellow of the Royal Society of Chemistry (RSC) and the Honorary Secretary of RSC London (North India Section). Apart from this, he is also a member of the American Chemical Society (ACS) and the faculty advisor of the International Student Chapter of ACS. After obtaining his PhD in 1986 from the University of Delhi, Prof. Sharma worked on Metal–Bimolecular interactions on a JSPS Post-Doctoral Fellowship at the University of Tokyo and the Kumamoto University. He has successfully supervised the research work of 36 PhD and MPhil students and published about 150 research as well as review articles in renowned international journals, including *Chemical Communications, Green Chemistry, Coordination Chemistry Reviews, ACS Sustainable Chemistry & Engineering, ACS Omega, Dalton Transactions* and many more. Recently, he has been nominated as the series advisor of the much prestigious *RSC Green Chemistry* Book Series and is also the ambassador of Bentham Publications. His research interests primarily focus on the fabrication of silica-based organic–inorganic hybrid materials for their applications as scavengers, sensors and catalysts, designing of novel metal-chelating inhibitors of transcription factor NF-κB-DNA binding, chemical speciation, molecular modeling studies, etc. Among several of his highly cited papers, one of his research articles on "An efficient copper-based magnetic nanocatalyst for the fixation of carbon dioxide at atmospheric pressure" won recognition in the form of one of the top cited 100 articles of *Nature Scientific Reports* in 2018. Also, his review article entitled "Silica-nanosphere-based organic–inorganic hybrid nanomaterials: Synthesis, functionalization and applications" in the field of catalysis was one of the highly downloaded articles of *Green Chemistry*. He has edited a book on *Hazardous Reagent Substitution* which is published by the Royal Society of Chemistry, London. Prof. Sharma is the distinguished recipient of several prestigious awards, including 2010 INSA-JSPS award to visit Japan, 2010 UGC-TEC award to visit Mauritius, 2002 INSA-JSPS award, 1999 World Green Award, 1998 Research Grant Award by RSC London, 1995 Indo-German Award and 1995 UGC National Research Scientist award. He is a UGC member of the expert committee for evaluation of national and international projects and DST WTI PAC committee member.

Contents

Chapter 2.   Silica Nanoparticles: A Smart and Promising Support Material for the Design of Heterogeneous Catalytic Systems

Sriparna Dutta, Rashmi Gaur, Shivani Sharma and Rakesh Kumar Sharma

Chapter 3.   Silica-Encapsulated Magnetic Nanoparticles

Yukti Monga, Radhika Gupta, Rashmi Gaur and Rakesh Kumar Sharma

Chapter 4.   Different Approaches for Surface Modification

Manavi Yadav, Yukti Monga, Gunjan Arora and Rakesh Kumar Sharma

**Chapter 5. Characterization of Metal-
Immobilized Silica Nanoparticles and
Silica-Coated Magnetic Nanoparticles   145**

*Rashmi Gaur, Shivani Sharma, Yukti Monga and
Rakesh Kumar Sharma*

**Chapter 6. Catalytic Applications of Silica-
Based Organic–Inorganic Hybrid Nanomaterials
for Different Organic Transformations   171**

*Radhika Gupta, Gunjan Arora, Manavi Yadav and
Rakesh Kumar Sharma*

**Chapter 7.   Other Potential Catalytic
Applications and Future Perspectives   221**

*Gunjan Arora, Sriparna Dutta, Radhika Gupta and
Rakesh Kumar Sharma*

Chapter 1

An Introduction to Silica-Based Organic–Inorganic Hybrid Nanostructured Catalytic Systems

Shivani Sharma, Manavi Yadav, Sriparna Dutta
and Rakesh Kumar Sharma*
*rksharmagreenchem@hotmail.com

1.1 Introduction

In today's modern era of globalization, a little word with big potential has rapidly insinuated itself into the world's consciousness. The word is "nano" which has conjured up speculation about a seismic shift in almost every aspect of day-to-day life, leading to undiscovered realms of science and creating new research domains. Visionaries tout it as the panacea for all our problems. In fact, nanotechnology has been considered to be the biggest engineering innovation that has led to tremendous enthusiasm among research groups of different scientific disciplines, such as physics, chemistry and biology (Figure 1.1) [1].

The idea of "nanotechnology" was first introduced in the year 1959 by Richard P. Feynman, one of the greatest Nobel Laureates, when he delivered a talk entitled *There's Plenty of Room at the Bottom* in the Annual Meeting of the American Physical Society at the California Institute of Technology [2]. This lecture, in which Feynman suggested the possibility of arranging the atoms in the way

---

*Corresponding author.

1

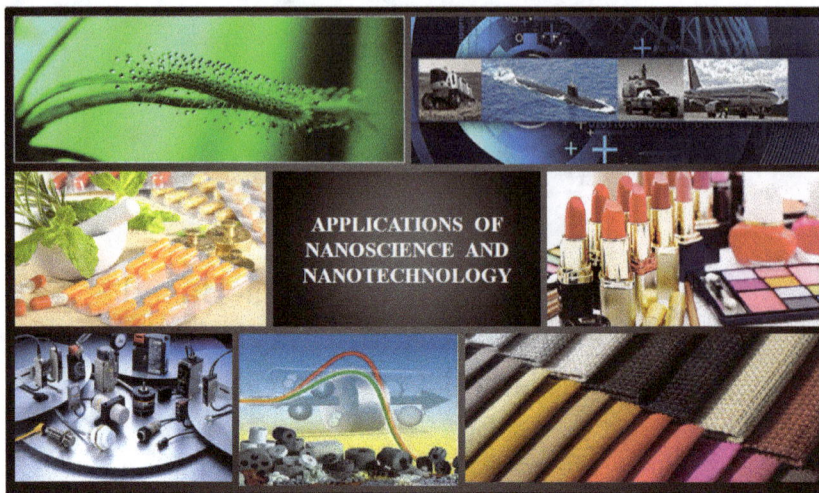

Figure 1.1. Diverse applications of nanoscience and nanotechnology.

we want, has become one of the classic talks of the 20th century. Inspired by Feynman's talk, K. Eric Drexler wrote a book entitled *Engines of Creation: The Coming Era of Nanotechnology* to theorize nanotechnology in depth and popularize the subject [3]. Since then, a lot of research on nanoscience and nanotechnology has been carried out throughout the globe, and it has resulted in the discovery of new types of materials that possess physical and chemical properties, which are not observed in their bulk counterparts. The prefix "nano" derived from the Greek word "dwarf" is used to refer to the length scale of one billionth of a meter (10^{-9}). It is interesting to note that an insect's compound eyes have tiny bumps that range in diameter from 50 to 300 nm (Figure 1.2). These bumps break up the cornea's even surface that cut down the glare which reflects off from their eyes, thereby helping the insect's camouflage.

In recent years, the design and development of organic–inorganic hybrid nanomaterials that synergistically integrate the functional organic moieties and inorganic building blocks into unique nanostructures have captivated the attention of several researchers working worldwide in the area of catalysis science as they enable environmentally friendly and benign catalytic processes [4]. Indeed,

Figure 1.2. Amazing ways animals use nanotechnology.

significant advancements have been made worldwide in the fabrication of organic–inorganic hybrid nanomaterials as active heterogeneous catalysts for several synthetic organic transformations [5]. These nanocatalysts have emerged as powerful tools for the efficient conversion of raw materials into useful chemicals of both industrial and pharmaceutical significance. Much efforts have been directed toward the design of quasi-homogeneous catalysts that incorporate silica-based nanoparticles (NPs) as support materials, owing to their outstanding inherent properties, such as large surface area, excellent mechanical and thermal stability, nanometer size and the presence of silanol groups on the surface, which allow the incorporation of a wide variety of functionalities [6]. Such precisely engineered catalytic systems that combine the best attributes of homogeneous as well as heterogeneous catalysts are free from diffusion problems generally associated with bulk catalytic systems, as these NPs can be dispersed in a wide range of organic solvents, eventually facilitating the accessibility of the substrates to the metal centers. Apart from this, these third-generation catalytic systems display high activity and selectivity, excellent stability, efficient recovery and recyclability (Figure 1.3).

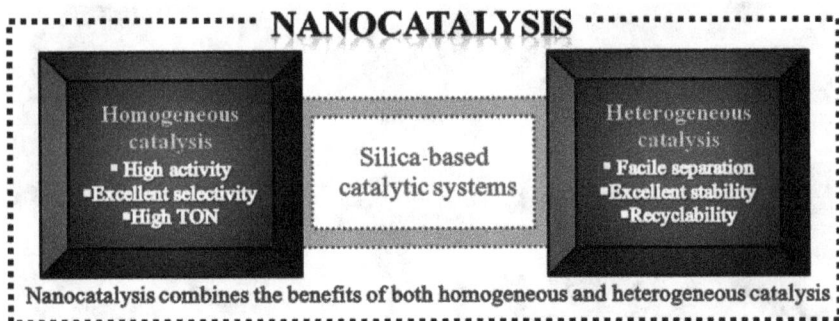

Figure 1.3. Advantages of nanocatalysis.

This chapter provides the readers a succinct overview of the introductory concepts related to nanomaterials and their significance predominantly in the field of catalysis. The ensuing sections shall focus on the strategies adopted for the development of organic–inorganic hybrid nanocatalytic systems, with special emphasis on the utilization of silica as a support material that are being used to expedite diverse organic transformations. The interesting case studies that throw light on how nanocatalysts are being successfully used for different chemical processes in industries form a special highlight of this chapter.

1.2 Nanomaterials — Definition and Classification

Nanomaterials can be defined as "any material that has an average particle size ranging from 1 to 100 nm, or in other words materials that have particles or constituents of nanoscale dimensions."

1.2.1 Classification of nanomaterials based on their origin

1.2.1.1 *Natural nanomaterials*

These materials exist in the natural world (animals and minerals) without human modification or processing and have remarkable properties because of their inherent nanostructures. Examples of some of the natural nanostructured materials include viruses, protein molecules, antibodies, minerals, such as clays, natural colloids such as milk and blood (liquid colloids), fog (aerosol type), gelatine (gel type), mineralized natural materials, such as shells, corals and bones, insect wings and opals, spider silk, lotus leaf, gecko feet, volcanic ash, and ocean spray [7].

1.2.1.2 *Artificial nanomaterials*

These are prepared deliberately through a well-defined mechanical and fabrication process. Examples of artificial nanomaterials include carbon nanotubes (CNTs) and semiconductor nanoparticles like quantum dots, etc. [8].

1.2.2 Classification of nanomaterials based on their dimensions (Figure 1.4)

1.2.2.1 *Zero-dimensional nanomaterials*

These nanomaterials have nanodimensions in all the three directions. The most common representation of zero-dimensional nanomaterials is *nanoparticles* which can

- be amorphous or crystalline,
- be single crystalline or polycrystalline,
- consist of single or multi-chemical elements,
- exhibit various shapes and forms,
- exist individually or be incorporated in a matrix,
- be metallic, ceramic or polymeric.

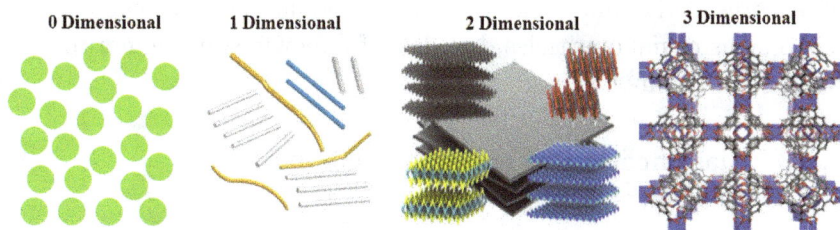

Figure 1.4. Types of nanomaterials based on their dimensions.

1.2.2.2 *One-dimensional nanomaterials*

These nanomaterials have one of the dimensions outside the nanometer range that leads to needle-shaped nanomaterials. Examples include nanotubes, nanorods and nanowires which can be

- amorphous or crystalline,
- single crystalline or polycrystalline,
- chemically pure or impure,
- embedded within another medium,
- metallic, ceramic or polymeric.

1.2.2.3 *Two-dimensional nanomaterials*

In these types of nanomaterials, two of the dimensions are not confined to the nanoscale that results in plate-like shapes. Examples include nanosheets, nanofilms, nanolayers and nanocoatings which can be

- amorphous or crystalline,
- made up of various chemical compositions,
- used as a single layer or multilayer structure,
- deposited on a substrate,
- integrated in a surrounding matrix material,
- metallic, ceramic or polymeric.

1.2.2.4 *Three-dimensional nanomaterials*

These are bulk materials that are characterized by having three arbitrary dimensions above 100 nm. With respect to the presence of features at the nanoscale, three-dimensional materials can contain dispersions of nanoparticles, bundles of nanowires and nanotubes as well as multi-nanolayers [9–12].

1.2.3 Classification of nanomaterials based on structural configurations

1.2.3.1 *Carbon-based nanomaterials*

These nanomaterials are made up of carbon and can be hollow spheres, ellipsoids or tubes. Spherical and ellipsoidal carbon nanomaterials are named as fullerenes, whereas cylindrical nanomaterials are defined as nanotubes. These carbon-based nanomaterials are applied in a wide range of applications, including improved films and coatings, electronics, sensing and so on.

1.2.3.2 *Metal-based materials*

As is clear from the name itself, the key component of the metal-based materials is metal. Such types of nanomaterials include quantum dots, nanogold, nanosilver and metal oxides such as titanium.

1.2.3.3 *Dendrimers*

Dendrimers are branched macromolecules having numerous chain ends that can be tailored to perform specific chemical functions, especially catalysis. Furthermore, three-dimensional dendrimers can be applied in drug delivery as well because of the presence of interior cavities into which other molecules can be placed.

1.2.3.4 *Composites*

Composites are multiphase solid materials formed by the combination of nanoparticles with other nanoparticles or with larger, bulk-type materials. Examples include colloids, gel copolymers and nano-sized clays [13, 14].

1.3 Synthetic Approaches of Nanomaterials

Primarily, there are two approaches to synthesize nanomaterials (Figure 1.5).

1.3.1 Bottom-up approach

In this approach, smaller building blocks such as atoms and molecules self-assemble together according to a natural physical principle or external driving force to generate larger nanostructures; or, in other words, instead of taking material away to make structures, the bottom-up approach selectively adds atoms to create nanostructures [15]. Typical examples include sol–gel processing [16], chemical

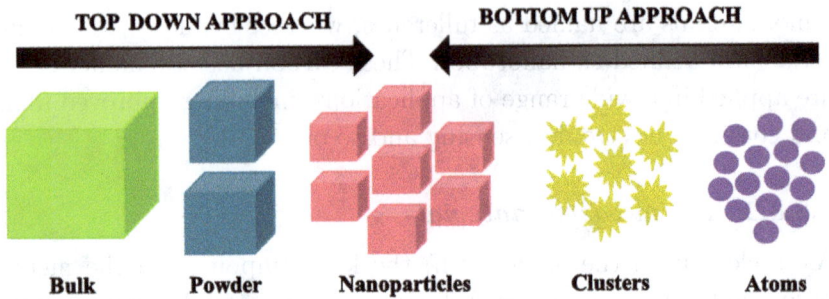

Figure 1.5. Top-down and bottom-up approaches for nanomaterial synthesis.

vapor deposition (CVD) [17], plasma- or flame-spraying synthesis [18], laser pyrolysis [19], and atomic or molecular condensation.

1.3.2   Top-down approach

As the name suggests, this approach essentially involves the breakdown of a larger system into correspondingly smaller structures. The top-down processing has been and will be the dominant process in semiconductor manufacturing. Examples include different kinds of lithographic techniques, such as cutting, electron beam, photo ion beam or X-ray lithograph cutting, etching, grinding, and ball milling [20–22].

1.4   Catalysis

1.4.1   History and background

The term "catalysis" was first coined by a Swedish scientist Jons Jakob Berzelius in the year 1835 to identify a new entity capable of promoting the occurrence of a chemical reaction [23]. It is important to note that catalysis can be traced back to ancient times if we consider, for instance, the fermentation processes, which are examples of biocatalysis. Catalysts had in fact already been used in few laboratories before that time, for example, the dehydrogenation of alcohol was carried out using metal catalysts by the Dutch chemist Martinus van Marum [24]. But, it was Berzelius who first recognized that the metal was not simply providing heat, but that it had a very

Figure 1.6. Schematic illustration of lowering of activation energy by catalyst.

special, unknown effect which he called the "catalytic force". Thirty years later, Michael Faraday demonstrated the ability of platinum to recombine hydrogen and oxygen [25]. Toward the end of the century, the Riga-born German chemist Wilhelm Ostwald discovered that catalyst is an agent that can only influence the rate of a reaction, not its final outcome, and itself remains unchanged [26]:

> *Catalysts actually don't produce reactions that are otherwise impossible but they simply assist already existing organic reactions. This is comparable to a walk in the mountains. The route to a neighboring valley may lead over a hill (Figure 1.6). But catalysts play the role of a local guide, who directs you through an old, unused railway tunnel, so that you can avoid a tiring climb. Thus, the guide makes your task of reaching the other valley convenient.*

Over the past few decades, catalysis has been playing a pivotal role in both academic research as well as industrial manufacturing sector with considerable potential of applications in everyday life, including fine chemicals, agrochemicals, pharmaceuticals, petroleum, polymers, electronics and so on. For instance, the catalytic converters in our cars are like small chemical plants that split noxious nitrogen oxides to form harmless nitrogen and oxygen molecules.

1.4.2 Performance dimensions

An ideal catalyst is expected to exhibit higher performance in the three dimensions explained in the following sections (Figure 1.7).

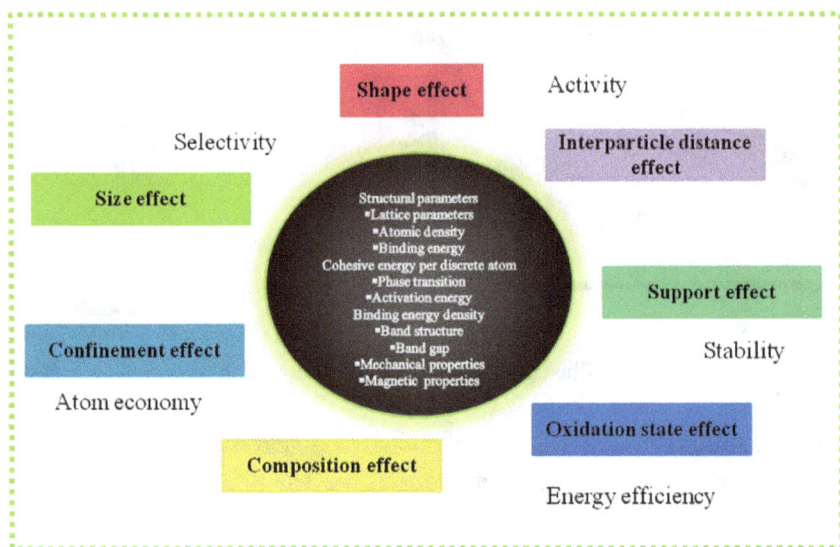

Figure 1.7. Factors that affect the catalytic activity.

1.4.2.1 *Selectivity*

Selectivity is the ability of a catalyst to direct a reaction to yield a particular product. In general, a selective catalyst is the one that would ideally lead to the formation of 100% of the desired product. In this manner, a raw material is converted more efficiently to the desired product without the generation of unnecessary waste while eliminating the cumbersome separation process.

1.4.2.2 *Activity*

It is the ability of a catalyst to increase the rate of reaction which depends upon the adsorption of reactants on the surface of the catalyst. The bond formed between the catalytic surface and the reactants must be strong enough to make the catalyst active, and simultaneously, it must not be so strong that the reactants get anchored on the catalyst surface, leaving no further space for new reactants. The measure for the activity is called as turnover frequency (TOF), which may be expressed as how many molecules of raw material are converted to product molecules by the catalyst per unit of time. Nanocatalytic systems result in high TOF values in comparison to their bulk counterparts.

1.4.2.3 *Lifetime of a catalyst*

The lifetime of a catalyst can also be called as catalyst durability. It is actually the total number of catalytic cycles the catalyst can undergo until it needs to be replaced. As a result, a highly durable catalyst allows the cost-effective synthesis of a desired compound before the process has to be interrupted for the replacement of the catalyst [27].

1.4.3 Factors that influence the performance dimensions

The three above-mentioned dimensions — "catalytic activity, selectivity and recyclability" — depend on the size, shape and surface composition of the nanocomposites [28]. Reduction in the size of the catalyst from bulk- to nano-level results in an increase in the surface area-to-volume ratio because of which the accessibility of catalytically active sites for substrate molecules increases, thereby leading to an overall increase in the TOF and selectivity. The size-dependent selectivity was shown by Hayashi *et al.* who investigated the catalytic activity of gold nanoparticles in the reaction of propylene with oxygen (O_2) and hydrogen (H_2) [29]. They observed that if the particle size was smaller than 2 nm, propane was formed instead of propylene oxide. The second influencing factor is the shape of nanoparticles that strongly affects the activity of a nanocatalyst. For instance, Narayanan *et al.* compared the ability of tetrahedral, cubic and "near-spherical" platinum (Pt) nanoparticles to catalyze the reaction of hexacyanoferrate (III) and thiosulfate ions and found that despite having similar size, the tetrahedral particle featuring a larger amount of edges and corners was the most active in comparison to the others [30].

Besides size and shape, surface composition also plays an important role in influencing the activity and selectivity of a catalyst. The effect of surface composition has been very well demonstrated by Pool and group who established that the catalytic activity of

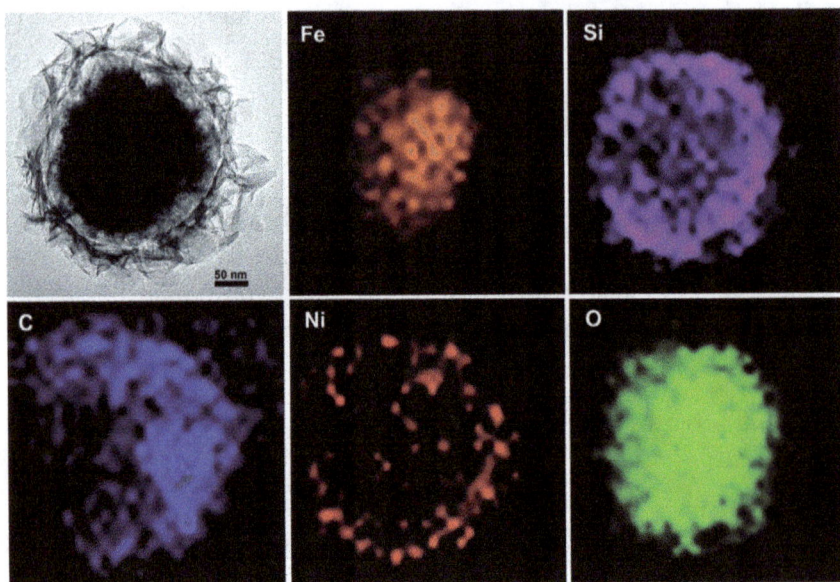

Figure 1.8. The elemental mapping shows homogeneous dispersion of Fe, Si, C, Ni and O elements in the $Fe_3O_4@SiO_2@Ni–L$ core–shell microspheres (adapted from Ref. [32]).

Co_{13} completely diminishes when a vanadium atom replaces one of the cobalt atoms in the composition [31]. Furthermore, surface composition also contributes to nanocatalyst durability and recoverability. It is a well-known fact that nanoparticles show a strong tendency to undergo agglomeration which results in deterioration in the catalytic activity. So, in order to increase the durability of a catalytic system, surface functionalization with suitable capping agents like polymers and surfactants is usually done. A prominent example is the protection of magnetic nanoparticles (MNPs) using silica as the coating agent which has been shown by Tan and co-workers (Figure 1.8) [32].

1.4.4 Heterogenization approach

Conventionally, catalysts are of two types: homogeneous and heterogeneous. Although homogeneous catalysis is considered to be one of the most proficient and environmentally sustainable routes to synthesize high added-value chemicals, the practical applications of many homogeneous catalysts in industrial processes are hindered by their high costs, multistep preparation and difficult separation from the reaction mixture. On the contrary, heterogeneous catalysis allows facile separation and recycling of the catalyst, yet it suffers from several drawbacks such as low activity and selectivity as compared to its homogeneous counterpart. Hence, the quest for novel recyclable catalysts that can bridge the gap between homogeneous and heterogeneous catalysis continues to be a challenging mission in the field of catalysis [33, 34]. Many synthetic strategies have been employed to surmount the challenges of catalyst separation which include nanofiltration and new separation techniques based on liquid–liquid phase separation, including ionic liquids (ILs), fluorous phases, supercritical solvents and polymeric supports. However, each of these enlisted methodologies suffers from serious drawbacks of efficiency or generation of secondary waste [35]. Recently, heterogenization of expensive but very active homogeneous metal complexes has become one of the most significant directions in the ongoing research in catalysis and a prominent example illustrating the efforts toward Green Chemistry [36]. Table 1.1 presents a comparative study between the three different categories of catalysts.

There have been extensive reports in the literature wherein various organic groups have been anchored onto support materials such as alumina, titania, silica, and zirconia using different immobilization techniques which are listed in the following sections.

1.4.4.1 *Covalent binding*

The covalent binding approach is utilized in order to build a strong bonding between the functional groups and the solid support materials (Figure 1.9) [37, 38]. Catalysts prepared using this strategy are devoid of the leaching problem and can be employed in a wide range of reaction conditions in comparison to other non-covalent interactions [39]. The most commonly used support matrix for the catalyst immobilization is silica which can be functionalized either by co-condensation (direct synthesis) or by post-grafting method [40, 41]. Direct synthesis of catalyst *via* the former method is done by sol–gel co-condensations of the siloxanes and organosiloxanes in the presence of structure-directing agents. In the latter approach,

Table 1.1.   Comparison between different types of catalysts.

Type of catalyst	Homogeneous catalysts	Heterogeneous catalysts	Heterogenized homogeneous catalysts
Examples	Acid–base catalysts and transition metal complexes	Metals, metal oxides and supported metal oxides	Immobilization of homogeneous catalyst on solid support
Phase	Liquid/gas	Solid	Solid
Active sites	Well-defined	Poorly defined	Well-defined
Concentration	Low	High	Low
Activity	High	Low	High
Selectivity	High	Low	High
Temperature	Low	High	Low, high
Thermal stability	Low	High	High
Product separation	Difficult	Easy	Easy
Catalyst regeneration	Generally problematic	Facile	Facile
Catalyst modification	Easy	Difficult	Easy
Catalyst cost	Low	High	High
Reaction rate	Slow	Fast	Fast
Reaction mechanism	Reasonably well understood	Poorly understood	Same as homogeneous catalysts
Lifetime of catalyst	Variable	Long	Long

modification of the silica surface is carried out *via* silylation, which is a chemical reaction between surface silanols (isolated, geminal or vicinal) and alkoxy- or chloro-organosilanes [42, 43]. The materials produced by post-grafting possess well-defined structures, while the co-condensation method produces materials having more uniform surface coverage of functional groups.

1.4.4.2 *Adsorption*

Another common immobilization technique is adsorption, in which active catalytic species are adsorbed onto the surface of a solid

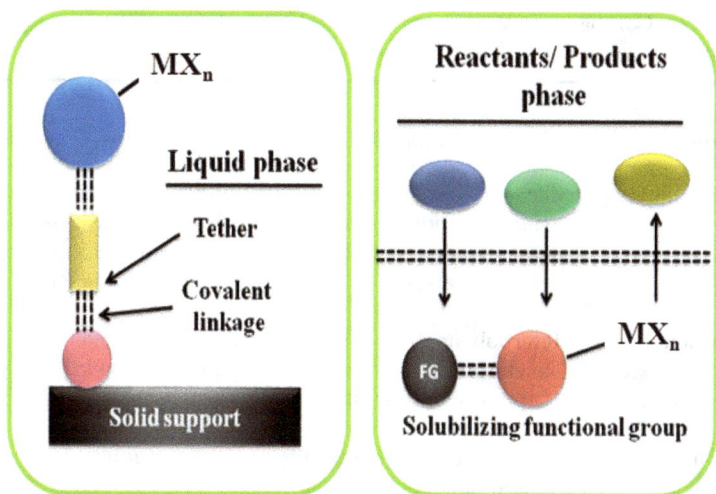

Figure 1.9. Approach depicting the covalent binding of metal complexes onto the surface of a solid support matrix.

support *via* physisorption [44, 45]. This method is relatively much easier in comparison to the covalent binding method, since it does not require modification of organic ligands for immobilization. However, catalysts synthesized by adsorption method could only be employed in limited organic reaction, solvents and temperatures because catalytic species tend to be leached into the reaction solution to some extent.

1.4.4.3 *Entrapment*

In this method, catalysts are assembled first and thereby entrapped inside the cage-like pores of porous materials, such as zeolites and MCM-41, like "ship in a bottle" [46, 47]. The dimension of the metal complexes/organic groups must be larger than the size of the pore mouth so as to prevent leaching of active catalytic species [48]. This synthetic strategy is not generally applied since it suffers from limitations on the diffusion of reactants and products as well as constraints on the interactions between substrates and active sites. Table 1.2 provides the comparison of various methodologies of immobilization of active catalytic sites onto a solid support matrix.

Table 1.2. Comparison of various methodologies of immobilization of active catalytic sites onto a solid support matrix.

Immobilization techniques	Examples	Pros/Cons
Covalent binding	Silica-supported propyl sulfonic acids; polymer resin-supported Co(III) Salen catalyst [41, 77]	Stable attachment, less leaching, increased complexity to synthesize the immobilization precursors
Adsorption	Heterpoly acids as anchoring agents between supports and metal complexes [45]	Limited reaction conditions prone to be leached
Entrapment	Polysaccharide-supported metal nanoparticles; Ni Salen complexes inside zeolites [47, 78]	Diffusion limitation

1.4.5   Use of metal oxides as support material for the design of heterogeneous catalysts

Metal oxides have emerged as attractive materials in diverse techno-logical applications, including microelectronics, sensors, piezoelectric devices, fuel cells, coating for the passivation of surfaces against corrosion and catalysis [49, 50]. In the emerging field of nanotech-nology, nanostructured materials such as ZnO, SiO_2, Fe_3O_4, TiO_2, and Al_2O_3 have been widely employed as solid support materials as they possess unique physicochemical properties with respect to bulk materials owing to their limited size and a high density of corner or edge surface sites (Figure 1.10).

1.4.6   Silica: An ultimate choice as a support material

Silica, or silicon dioxide (SiO_2), which occurs either in crystalline or amorphous forms, is one of the most abundant components of the earth's crust found in nature. It has a three-dimensional network structure, built up by packing $[SiO_4]$ units, and a bulk structure that terminates at the surface in two different ways, that is, with oxygen

Synthesis of solid supported organic-inorganic hybrid nanostructured catalytic systems

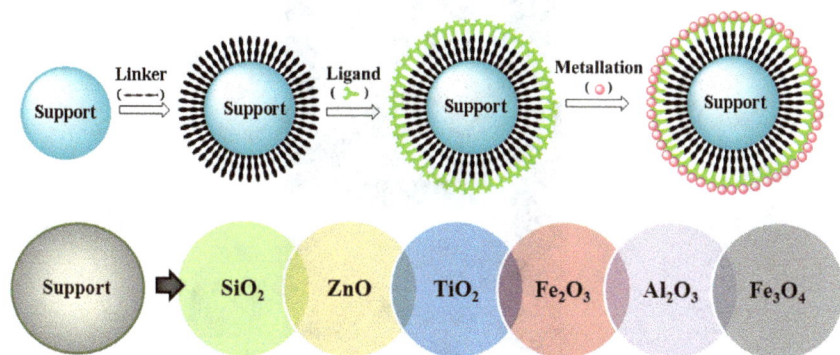

Different nanomaterials used in literature as solid support matrix

Figure 1.10. Schematic illustration depicting the fabrication of solid-supported organic–inorganic hybrid nanostructured catalytic systems.

on the surface through siloxane groups (μSi–O–Siμ) or silanol groups (μSi–OH) [51]. There are three types of silanol groups, namely, vicinal (hydrogen-bonded silanols), geminal (two silanol groups attached to the same silicon atom) or isolated (no hydrogen bonds possible) silanol sites, which shall be discussed in more detail in Chapter 2.

Recently, there has been an upsurge of interest in the utilization of silica nanoparticles (SNPs) as a solid support matrix for the design and development of third-generation, precisely engineered heterogeneous nanostructured catalytic systems [52, 53]. This excellent catalytic activity of SNP-based catalysts is attributed to their intrinsic, unique and adjustable physicochemical properties, such as high surface area, nanometer size and rigid framework, with excellent chemical, thermal and mechanical stability (Figure 1.11).

An additional feature that multiplies their use manifold as an exceptional support material is the possibility to functionalize both the exterior and interior pore surfaces with several organic moieties [54]. Moreover, these NPs can be easily dispersed in several solvents which facilitates the accessibility of the reactants to the catalytic active sites, thereby avoiding the diffusion problems

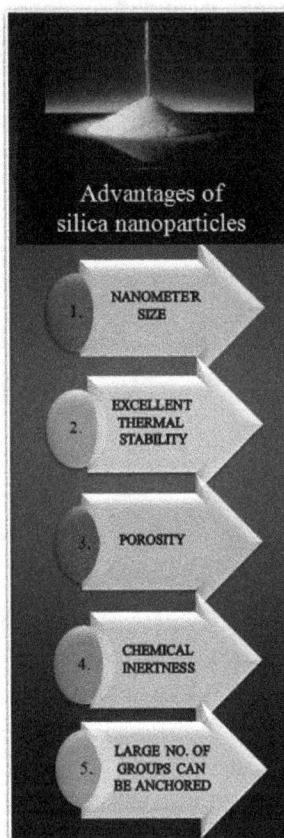

Figure 1.11. Advantages of SNPs.

associated with heterogeneous bulk catalytic systems [55, 56]. The functionalization of silica-based materials with hybrid atoms, NPs and organic groups has been a long-term research focus, owing to their ability to tune the inherent inactive nature of silica and broaden its applicability, especially in the field of catalysis.

To date, several studies have been directed toward the rational design of the size, shape and composition of the active SNPs in order to develop highly active and selective nanocatalysts [57–61]. For example, different research groups worldwide have fabricated nanosilica having different morphologies, as shown in Figure 1.12.

Figure 1.12. Different morphologies of SNPs reported in literature. Transmission electron microscopy (TEM) images of (a) fibrous SiO_2 nanospheres KCC-1, (b) hollow mesoporous SiO_2 NPs, (c) calcined mesoporous silica nanotubes, (e) SiO_2 nanospheres, and SEM images of (d) jellyfish-like nanowires, (f) right-handed helical silica nanotube with mesopores (adapted from Refs. [57–61]).

Catalyst separation from reaction mixture remains a challenging issue for almost every synthetic manufacturing process. Conventional separation techniques such as filtration are not efficient enough because of the nanodimensions of the catalyst particles. This limitation seriously hinders the economics and sustainability of this nanocatalyst. To overcome this issue, core–shell-structured silica-encapsulated MNPs have emerged as a viable solution: the insoluble and paramagnetic nature of the core material enables easy and efficient separation of the catalyst from the reaction mixture with an external magnet [62, 63]. Magnetic separation generally offers high efficiency and specificity when compared with equivalent centrifugation or filtration methods (Figure 1.13). Considering the immense benefits of MNPs (i.e. excellent separation properties and low toxicity coupled with better biocompatibility), several research groups have realized their potential as a versatile solid support for the

SILICA-AN ULTIMATE CHOICE
Paving the way to Magnetically recyclable core-shell nanocatalysts

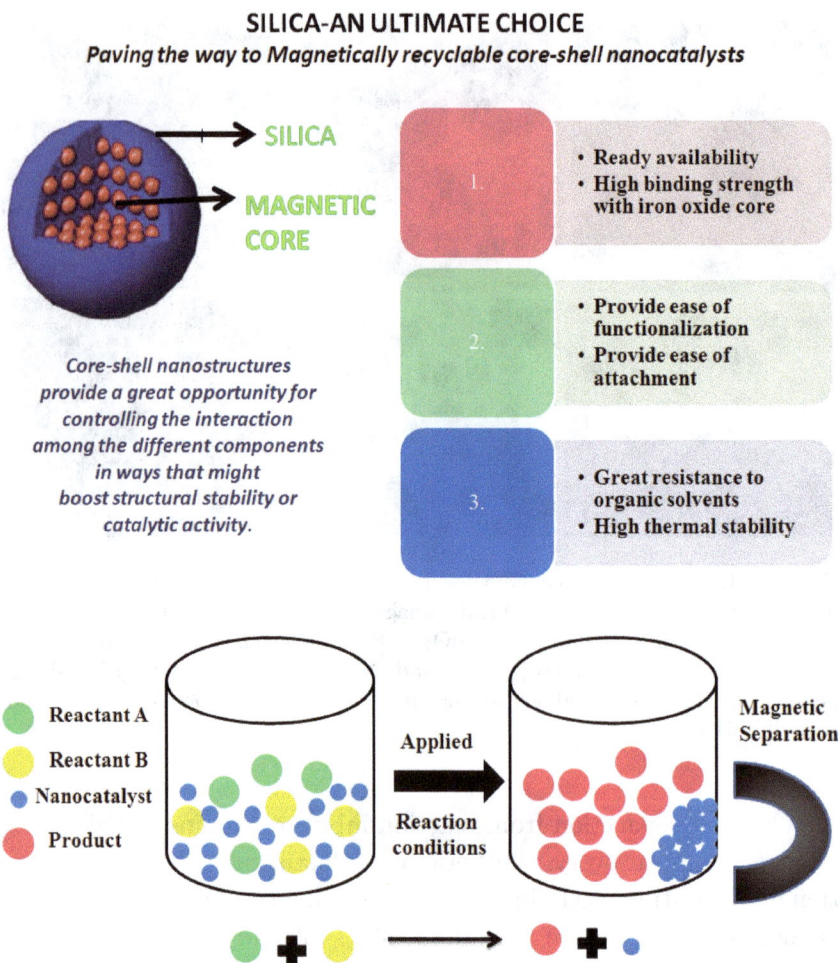

SILICA

MAGNETIC CORE

1.
- Ready availability
- High binding strength with iron oxide core

2.
- Provide ease of functionalization
- Provide ease of attachment

3.
- Great resistance to organic solvents
- High thermal stability

Core-shell nanostructures provide a great opportunity for controlling the interaction among the different components in ways that might boost structural stability or catalytic activity.

Reactant A
Reactant B
Nanocatalyst
Product

Applied
Reaction conditions

Magnetic Separation

Figure 1.13.   Advantages of silica-encapsulated MNPs.

fabrication of novel hybrid catalytic systems which exhibit enhanced activity and selectivity. For instance, Sadeghzadeh fabricated a heteropolyacid containing an IL-based organosilica (Fe_3O_4/KCC-1/ IL/HPW) catalyst and subsequently utilized it for the synthesis of tetrahydrodipyrazolopyridines [64]. The catalyst could be recovered and reused at least 10 times with no decrease in its activity and selectivity. Further, all such literature reports wherein either silica

NPs or silica-encapsulated Fe_3O_4 NPs have been utilized as support material for catalyst synthesis have been included in detail in the subsequent chapters.

1.5   A Paradigm Shift Toward Green Chemistry and Nanocatalysis

Nanostructured catalytic systems play a significant role in dealing with the challenges of energy and sustainability. With an aim to avoid the use of toxic reagents, volatile organic solvents, hazardous reaction conditions as well as challenging and time-consuming wasteful separations, and recently, the design and development of greener and environmentally benign catalysts have drawn significant interest of researchers and chemists worldwide [65–69]. In fact, nanocatalysis adds greenness to the chemical process by the implementation of 13 out of 24 principles of Green Chemistry and Engineering which have been discussed briefly in the following:

(1) Higher selectivity achieved by nanocatalysts as compared to the conventional catalytic systems allows one to perform chemical reactions in a specific manner with the least possible consumption of substances. This is in accordance with the first principle of Green Chemistry and second principle of Green Engineering: *It is better to prevent waste than to treat or clean up waste after it is formed.* For example, hydrogenation of cyclohexanone in the presence of nanocatalysts helps in accomplishing the generation of by-products, thereby leading to increased selectivity.

(2) Some of the new nanocatalytic processes utilize water instead of toxic organic solvents. This is an example of the third principle of Green Chemistry: *Wherever practicable, synthetic methodologies should be designed to use and generate substances that possess little or no toxicity to human health and the environment.*

(3) According to the fourth principle of Green Chemistry as well as Green Engineering, *Chemical products should be designed to preserve efficacy of function while reducing toxicity.* This is well

illustrated by the synthesis of non-toxic ZnO nanocatalyst or by the stabilization of gold nanoparticles using tea extract that contains polyphenols.

(4) The fifth principle of Green Chemistry, *The use of auxiliary substances (solvents, separation agents, etc.) should be made unnecessary wherever possible and innocuous when used,* and the third principle of Green Engineering, *Separation and purification operations should be designed to minimize energy consumption and materials use,* are well exemplified by the use of magnetic nanocatalysts that helps in avoiding challenging and time-consuming product purification and separation process.

(5) Gao [79] reported the hydrolysis of esters by employing nanocatalyst under mild reaction conditions which follows the sixth principle: *Synthetic methods should be conducted at ambient temperature and pressure.*

(6) The eighth principle suggests: *Unnecessary derivatization (blocking group, protection/deprotection, temporary modification of physical/chemical processes) should be avoided whenever possible.* For example, Parasher *et al.* [80] developed a nanostructured catalyst anchored onto acid-functionalized solid support for catalyzing the direct synthesis of H_2O_2 that was unachievable using traditional methods.

(7) The introduction of a nano-sized zeolite catalyst for the well-known Friedel–Crafts demonstrates the practicability of the ninth principle: *Catalytic reagents (as selective as possible) are superior to stoichiometric reagents.*

(8) Several studies have been carried out, wherein nanocatalysts have made an important contribution toward ensuring safer chemical processes for organic molecules that fit the last principle of Green Chemistry, *Substances and the form of a substance used in a chemical process should be chosen so as to minimize the potential for chemical accidents, including releases, explosions, and fires,* and the first principle of Green Engineering, *Designers need to strive to ensure that all material and energy inputs and outputs are as inherently nonhazardous as possible.*

(9) An improved naphtha hydrogenation synthetic route using the reforming nanocatalyst reported by Zhou *et al.* [61] corresponds

to the fourth principle of Green Engineering that reads as follows: *Products, processes, and systems should be designed to maximize mass, energy, space, and time efficiency.*

(10) The surface of CNTs can be tailored with different functionalizing agents/linkers in order to achieve the desired catalytic activity. This example demonstrates the implementation of the eighth principle of Green Engineering: *Design for unnecessary capacity or capability (e.g. "one size fits all") solutions should be considered a design flaw.*

1.6 Nanocatalysis in Chemical Industries

Nanomaterials are being employed extensively as efficient catalytic materials in a wide range of industrial segments spanning refinery, petrochemical, pharmaceuticals, chemical, food processing sectors and others (Figure 1.14). According to the Global Industry Analysis, global nanocatalyst market has already reached USD 2,900. Not surprisingly, nanocatalysis is a growing business. The list of companies that have already patented and/or commercialized technologies relating to nanocatalysts is already impressive. The dominant global players include Argonide Corporation, BASF Catalyst LLC, BASF SE, Bayer AG, Catalytic Solution, Inc., Evonik Degussa GmbH, Genencor International, Inc., Headwaters Nanokinetix, Inc., Hyperion Catalysis International Inc., Johnson Matthey PLC, MACH I, Inc., Nanophase Technologies Corporation, NanoScale Corporation, NexTech Materials Ltd., Oxonica, PQ Corporation, Sachtleben Chemie Gmbh, Sud-Chemie AG, Umicore NV and Zeolyst International, among others [70, 71]. Section 1.7 includes a few case studies related to the development and commercialization of catalysts by some of the renowned international companies.

1.7 Case Studies

1.7.1 Johnson Matthey PLC

Johnson Matthey formerly known as Johnson & Cock, a British multinational speciality chemical company headquartered in the United Kingdom, is a global leader in sustainable technologies.

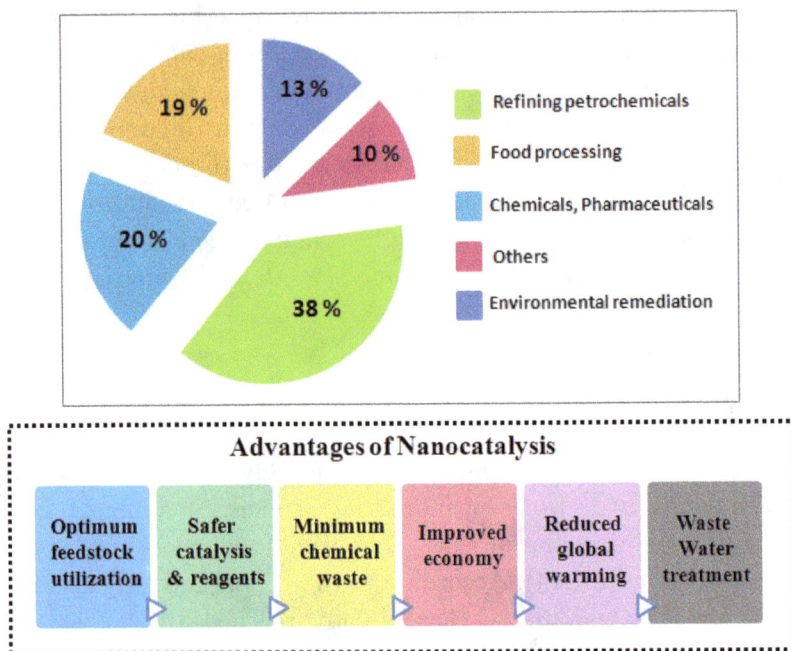

Figure 1.14.   Nanocatalysis in chemical industries.

This company primarily operates in five divisions: Emission Control Technologies, Process Technologies, Precious Metal Products, Fine Chemicals and New Businesses. In particular, the Emission Control Technologies division developed and commercialized novel catalytic systems that helped in controlling harmful emissions from heavy- and light-duty vehicles for which the company won a *MacRobert Award* in the year 2000. In fact, today, approximately one-third of all cars worldwide are built with a Johnson Matthey catalyst [72]. Recently, this company investigated the effect of hydrogen over platinum and combined platinum–palladium diesel oxidation catalysts (DOCs) (Pt and Pt–Pd DOCs) on the oxidation of CO, HC and NO. The results of their study have shown that even a small amount of H_2 has a positive effect on CO and HC oxidation as well as NO_2 formation over both Pt and Pt–Pd DOCs which might be due to either the increase in catalyst temperature caused by H_2 oxidation or the increase in rate of reaction because of accessibility of catalytic active sites. Also, a range of nickel and rhodium catalysts were evaluated for the conversion of model tar compounds (toluene, naphthalene or benzene) under real conditions using oak-derived syngas from National Renewable Energy Laboratory Pilot Scale Gasification unit. It was observed that Rh-based catalysts showed better hydrocarbon conversion in comparison to Ni catalysts at comparable conditions. The Johnson Matthey team have also fabricated a novel Al_2O_3-supported NiO nanocatalyst of 3 mm size in order to improve syn gas quality by increasing the amount of lighter fractions of H_2 and CO and reducing heavier fractions of CH_4 and CO.

1.7.2 BASF Catalyst LLC

BASF, a German chemical company, is the largest chemical producer in the world which comprises subsidiaries and joint ventures in more than 80 countries and operates six integrated production sites [73]. With the increasingly stringent environmental laws, BASF has been finding different ways to enhance catalyst performance in industrial organic transformations. Working in the area of nanocatalysis, this company came up with *NanoSelect LF 100* and *NanoSelect LF 200* which are the world's first lead-free alternatives to Lindlar catalysts that have long been used in alkyne to alkene hydrogenation. These two catalysts which are built on NanoSelect Platform differ in the support material utilized; the support material in case of LF 100 is activated carbon while the support material is alumina silicate powder in case of LF 200. In comparison to the conventional Lindlar catalysts, *LF 100 and 200* require less Pd *(Pd content of Lindlar catalysts is about 5% by weight, and Pd content for LF 100 and 200 is around 0.6% Pd by weight)* to achieve the same activity which results in enormous cost reduction in the hydrogenation process. In recognition of its NanoSelect catalyst technology platform, BASF won the *Green Excellent Award* from Frost & Sullivan in August 2009.

1.7.3 Hyperion Catalysis International Inc.

Found in the year 1982, Hyperion Catalysis International aims to develop novel forms and morphologies of carbon which find application in diverse fields. Over the years, it has emerged as a world leader in CNT development and commercialization [74]. Recently, Ma and group from Hyperion Catalyst International have developed and utilized CNTs as a support material for a variety of catalysts, including noble metals, transition metals and transition metal carbides, nitrides or derivatives. They compared the catalytic activities of CNT-supported Pd catalysts with the commercial Pd catalyst supported on activated carbon and found that with the same loading and particle sizes, the former catalyst showed nearly doubled activity and better selectivity in cyclohexene hydrogenation of carbon. The reason behind the enhanced catalytic performance of the fabricated CNT-supported Pd catalysts could be attributed to a possible electronic effect-induced catalyst support interaction.

1.7.4 Nanostellar

Platinum is the most expensive component of DOCs that are needed to meet the strict emission regulations for numerous light-duty and heavy-duty diesel vehicles produced annually worldwide. Based on platinum or palladium alloy, Nanostellar, located at Silicon Valley, California which is a start-up developer of nanostructured catalysts, introduced the first-generation catalytic system in mid-2006 that showed 25–30% higher performance in comparison to commercial pure Pt catalysts for diesel vehicle emission control. Thereafter, Nanostellar's second-generation product, i.e. NS gold (TM) catalyst, which is a trimetal formulation of Au, Pt and Pd, was manufactured that could reduce diesel hydrocarbon emission by as much as 40% more as compared to the other commercial catalytic systems [75].

1.7.5 The Nano Science and Technology Institute (NSTI)

As a result of the merger between different scientific societies, Nano Science and Technology Institute (NSTI) was created in 1997 and

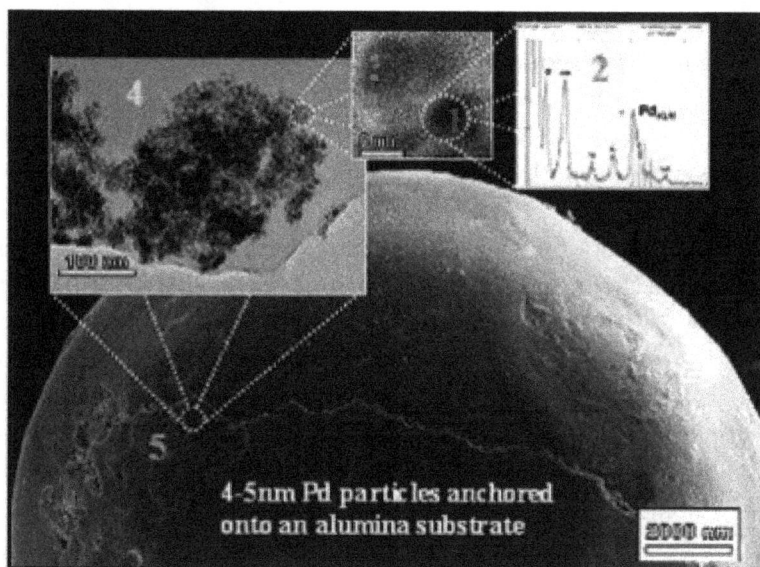

Figure 1.15. TEM and SEM of the nano-Pd–Pt catalyst developed by HTI company (HTI website) (adapted from Ref. [76]).

headquartered in Austin, Texas with office in Cambridge, Massachusetts and Danvilli, California. This Institute aims to promote nano and other recent technologies through education, conventions, business publishing and research service. Recently, NSTI came up with a breakthrough nanocatalyst called NxCAT^{TM} that enables the synthesis of hydrogen peroxide directly from its elements, hydrogen and oxygen. The NxCAT^{TM} catalyst is a palladium–platinum catalyst supported on spherical Al_2O_3 having a uniform size of 5 nm (Figure 1.15) that not only eliminates hazardous reaction conditions and chemicals of the existing process but also reduces both energy utilization as well as capital costs by almost 50% [76]. It is noteworthy that this catalytic process involves the use of innocuous, renewable feedstock and generates no toxic waste at all.

1.8 Conclusion

The grand challenge facing the chemical and allied industries in the 21st century is the development of green and sustainable materials for diverse synthetic processes that lead to the generation of value-added products. Thus, there is a noticeable paradigm shift toward cleaner manufacturing. Nanotechnology-based processes have enabled the development of economic and environmentally benign synthetic pathways to produce high-performance catalytic nanomaterials. These nanostructured materials have emerged as powerful catalysts for the efficient transformation of raw materials into valuable chemicals and fuels. There are extensive reports wherein SNPs and silica-encapsulated magnetic NPs have been employed as solid supports for the fabrication of organic–inorganic hybrid nanocatalysts, possessing several outstanding features, such as high activity and selectivity, excellent stability, efficient recovery and recyclability. The prime objective of this chapter is to broaden the outlook of the readers about the research progress on silica-based organic–inorganic hybrid materials and their applications as catalysts. To do so, an introduction of nanomaterials followed by a detailed discussion on their applications as potent catalytic materials along with different fabrication techniques have been presented. Also, a few case studies have been discussed, which summarize the substantial achievements made by several renowned companies, such as Johnson Matthey, BASF, Hyperion Catalysis International Inc., Nanostellar, and MACH I, Inc. in the field of nanocatalysis.

References

[1] C. R. Kagan, L. E. Fernandez, Y. Gogotsi, P. T. Hammond, M. C. Hersam, A. E. Nel, P. S. Weiss, *ACS Nano* **2016**, *10*, 9093–9103.
[2] R. Feynman, *Miniaturization*, H. D. Gilbert (ed.), Reinhold, New York, **1961**.
[3] R. Baum, *Chemical & Engineering News* **2003**, *81*, 37–42.
[4] R. K. Sharma, S. Sharma, S. Dutta, R. Zboril, M. B. Gawande, *Green Chemistry* **2015**, *17*, 3207–3230.
[5] L. L. Chng, N. Erathodiyil, J. Y. Ying, *Accounts of Chemical Research* **2013**, *46*, 1825–1837.
[6] T. Suteewong, H. Sai, J. Lee, M. Bradbury, T. Hyeon, S. M. Gruner, U. Wiesner, *Journal of Materials Chemistry* **2010**, *20*, 7807–7814.
[7] C. Buzea, I. I. Pacheco, K. Robbie, *Biointerphases* **2007**, *2*, 17–71.
[8] A. Fabbro, S. Bosi, L. Ballerini, M. Prato, *ACS Chemical Neuroscience* **2012**, *3*, 611–618.

[9] J. N. Tiwari, R. N. Tiwari, K. S. Kim, *Progress in Materials Science* **2012**, *57*, 724–803.

[10] V. V. Pokropivny, V. V. Skorokhod, *Materials Science and Engineering C.* **2007**, *27*, 990–993.

[11] J. Yuan, A. H. Müller, *Polymer* **2010**, *51*, 4015–4036.

[12] B. K. Teo, X. H. Sun, *The Journal of Cluster Science* **2007**, *18*, 346–357.

[13] S. Bhatia, *Nanoparticles Types, Classification, Characterization, Fabrication Methods and Drug Delivery Applications: Natural Polymer Drug Delivery Systems.* Springer International Publishing, Cham, **2016**, pp. 33–93.

[14] I. Y. Jeon, J. B. Baek, *Materials* **2010**, *3*, 3654–3674.

[15] X. Wang, J. Zhuang, Q. Peng, Y. Li, *Nature* **2005**, *437*, 121.

[16] M. Epifani, C. Giannini, L. Tapfer, L. Vasanelli, *The Journal of the American Ceramic Society* **2000**, *83*, 2385–2393.

[17] M. Adachi, S. Tsukui, K. Okuyama, *The Journal of Nanoparticle Research* **2003**, *5*, 31–37.

[18] U. R. Kortshagen, R. M. Sankaran, R. N. Pereira, S. L. Girshick, J. J. Wu, E. S. Aydil, *Chemical Reviews* **2016**, *116*, 11061–11127.

[19] S. Veintemillas-Verdaguer, M. P. Morales, C. J. Serna, *Materials Letters* **1998**, *35*, 227–231.

[20] T. P. Yadav, R. M. Yadav, D. P. Singh, *Nanoscience and Nanotechnology*, **2012**, *2*, 22–48.

[21] Y. Li, J. Scott, Y. T. Chen, L. Guo, M. Zhao, X. Wang, W. Lu, *Materials Chemistry and Physics* **2015**, *162*, 671–676.

[22] J. F. De Carvalho, S. N. De Medeiros, M. A. Morales, A. L. Dantas, A. S. Carriço, *Applied Surface Science* **2013**, *275*, 84–87.

[23] R. Winderlich, *The Journal of Chemical Education* **1948**, *25*, 500.

[24] A. J. B. Robertson, *Platinum Metals Review* **1975**, *19*, 64–69.

[25] C. Doornkamp, V. Ponec, *Journal of Molecular Catalysis A: Chemical* **2000**, *162*, 19–32.

[26] G. Ertl, *Angewandte Chemie International Edition* **2009**, *48*, 6600–6606.

[27] S. Olveira, S. P. Forster, S. Seeger, *Journal of Nanotechnolgy* **2014**, 1–19.

[28] S. Cao, F. F. Tao, Y. Tang, Y. Li, J. Yu, *Chemical Society Reviews* **2016**, *45*, 4747–4765.

[29] T. Hayashi, K. Tanaka, M. Haruta, *Journal of Catalysis* **1998**, *178*, 566–575.

[30] R. Narayanan, M. A. El-Sayed, *The Journal of Physical Chemistry B* **2007**, *107*, 12416–12424.

[31] R. Pool, *Science* **1990**, *248*, 1186–1188.

[32] L. Tan, X. Zhang, Q. Liu, J. Wang, Y. Sun, X. Jing, J. Liu, D. Songa, L. Liuc, *Dalton Transactions* **2015**, *44*, 6909–6917.

[33] L. Falivene, S. M. Kozlov, L. Cavallo, *ACS Catalysis* **2018**, *8*, 5637–5656.

[34] C. Coperet, M. Chabanas, R. Petroff Saint-Arroman, J. M. Basset, *Angewandte Chemie International Edition* **2003**, *42*, 156–181.

[35] D. J. Cole-Hamilton, *Science* **2003**, *299*, 1702–1706.

[36] A. E. Collis, I. T. Horvath, *Catalysis Science & Technology* **2011**, *1*, 912–919.

[37] T. Shimada, K. Aoki, Y. Shinoda, T. Nakamura, N. Tokunaga, S. Inagaki, T. Hayashi, *Journal of the American Chemical Society* **2003**, *125*, 4688–4689.

[38] P. K. Jal, S. Patel, B. K. Mishra, *Talanta* **2004**, *62*, 1005–1028.

[39] H. Liu, X. Xue, T. Li, J. Wang, W. Xu, M. Liu, Y. Wu, *RSC Advances* **2016**, *6*, 84815–84824.

[40] A. Walcarius, C. Delacôte, *Chemistry of Materials* **2003**, *15*, 4181–4192.

[41] W. Long, C. W. Jones, *ACS Catalysis* **2011**, *1*, 674–681.

[42] P. V. Der Voort, E. F. Vansant, *Journal of Liquid Chromatography & Related Technologies* **1996**, *19*, 2723–2752.

[43] D. Rother, T. Sen, D. East, I. J. Bruce, *Nanomedicine* **2011**, *6*, 281–300.

[44] L. Jiao, J. R. Regalbuto, *Journal of Catalysis* **2008**, *260*, 329–341.

[45] R. L. Augustine, S. K. Tanielyan, N. Mahata, Y. Gao, A. Zsigmond, H. Yang, *Applied Catalysis A: General* **2003**, *256*, 69–76.

[46] R. A. Sheldon, M. Wallau, I. W. Arends, U. Schuchardt, *Accounts of Chemical Research* **1998**, *31*, 485–493.

[47] D. Chatterjee, H. C. Bajaj, A. Das, K. Bhatt, *Journal of Molecular Catalysis* **1994**, *92*, L235–L238.

[48] C. Perego, R. Millini, *Chemical Society Reviews* **2013**, *42*, 3956–3976.

[49] H. J. Freund, G. Pacchioni, *Chemical Society Reviews* **2008**, *37*, 2224–2242.

[50] A. Harriman, I. J. Pickering, J. M. Thomas, P. A. Christensen, *The Journal of the Chemical Society, Faraday Transactions* **1988**, *84*, 2795–2806.

[51] I. I. Slowing, J. L. Vivero-Escoto, B. G. Trewyn, V. S. Y. Lin, *Journal of Materials Chemistry* **2010**, *20*, 7924–7937.

[52] A. Saad, C. Vard, M. Abderrabba, M. M. Chehimi, *Langmuir* **2017**, *33*, 7137–7146.

[53] A. Corma, H. Garcia, *Advanced Synthesis & Catalysis* **2006**, *348*, 1391–1412.

[54] J. Kecht, A. Schlossbauer, T. Bein, *Chemistry of Materials* **2008**, *20*, 7207–7214.

[55] Z. S. Qureshi, P. B. Sarawade, M. Albert, V. D'Elia, M. N. Hedhili, K. Köhler, J. M. Basset, *ChemCatChem.* **2015**, *7*, 635–642.

[56] J. Shabir, C. Garkoti, D. Sah, S. Mozumdar, *Catalysis Letters* **2018**, *148*, 194–204.

[57] H. Liu, Z. Huang, J. Huang, S. Xu, M. Fang, Y. G. Liu, X. Wu, S. Zhang, *Scientific Reports* **2016**, *6*, 22459.

[58] V. Polshettiwar, D. Cha, X. Zhang, J. M. Basset, *Angewandte Chemie International Edition* **2010**, *49*, 9652–9656.

[59] Y. Yu, H. Qiu, X. Wu, H. Li, Y. Li, Y. Sakamoto, Y. Inoue, K. Sakamoto, O. Terasaki, S. Che, *Advanced Functional Materials* **2008**, *18*, 541–550.

[60] Y. Yang, M. Suzuki, H. Fukui, H. Shirai, K. Hanabusa, *Chemistry of Materials* **2006**, *18*, 1324–1329.

[61] B. Zhou, H. Trevino, Z. Wu, Z. Zhou, C. Liu, US patent no. 7569508, Headwaters Technology Innovation LLC, 2005.

[62] Q. Zhang, I. Lee, J. B. Joo, F. Zaera, Y. Yin, *Accounts of Chemical Research* **2012**, *46*, 1816–1824.

[63] S. Dutta, S. Sharma, A. Sharma, R. K. Sharma, *ACS Omega* **2017**, *2*, 2778–2791.

[64] S. M. Sadeghzadeh, *RSC Adv.* **2016**, *6*, 75973–75980.

[65] I. T. Horváth, P. T. Anastas, *Chemical Reviews* **2007**, *107*, 2169–2173.

[66] S. B. Kalidindi, B. R. Jagirdar, *ChemSusChem* **2012**, *5*, 65–75.

[67] R. Narayanan, *Green Chemistry Letters and Reviews* **2012**, *5*, 707–725.

[68] V. Polshettiwar, R. S. Varma, *Green Chemistry* **2010**, *12*, 743–754.

[69] Y. Liu, G. Zhao, D. Wang, Y. Li, *National Science Review* **2015**, *2*, 150–166.

[70] G. Hutchings, *Nanocatalysis: Synthesis and Applications*, John Wiley & Sons, **2013**.

[71] V. Polshettiwar, *Angewandte Chemie International Edition* **2013**, *52*, 11199–11199.

[72] B. T. Johnson, *Platinum Metals Review* **2008**, *52*, 23–37.

[73] S. Jenkins, *Chemical Engineering* **2010**, *117*, 17.

[74] J. Ma, J. Yang, D. Moy, R. Hoch, *Journal of the American Chemical Society* **1994**, *116*, 7935–7936.

[75] A. Wittstock, J. Biener, J. Erlebacher, M. Bäumer, *Nanoporous Gold: From an Ancient Technology to a High-Tech Material*, Royal Society of Chemistry, **2012**.

[76] https://www.epa.gov/greenchemistry/presidential-green-chemistry-challenge-2007-greener-reaction-conditions-award.

[77] D. A. Annis, E. N. Jacobsen, *Journal of the American Chemical Society* **1999**, *121(17)*, 4147–4154.

[78] M. Zeng, C. Qi, J. Yang, B. Wang, X. Zhang, *Industrial Engineering Chemistry Research* **2014**, *53(24)*, 10041–10050.

[79] Y. Gao, Nano-reagents with cooperative catalysis and their uses in multiple phase reactions, US patent no. 7951744, Southern Illinois University, Carbondale, Ill, USA, **2007**.

[80] S. Parasher, M. Rueter, and B. Zhou, Nanocatalyst ancored onto acid functionalized solid support and method of making and using same, US patent no. 7045481, Headwaters Nanokinetix, Inc., Lawrenceville, NJ, USA, **2005**.

Chapter 2

Silica Nanoparticles: A Smart and Promising Support Material for the Design of Heterogeneous Catalytic Systems

Sriparna Dutta, Rashmi Gaur, Shivani Sharma
and Rakesh Kumar Sharma*
*rksharmagreenchem@hotmail.com

2.1 Introduction

The past decades have witnessed a surge in the research interest of the academicians as well as industrialists toward the development of well-defined nanostructured materials possessing controllable size, morphology and geometry [1–5]. As evident, once the particle size comes down to the nano-level, there is a drastic change in their properties, which can be fruitfully utilized for different applications. Particularly focusing on catalysis, *a research area that has driven the development of society*, one finds that, with the advent of nanotechnology, a net economic boom has been observed which is obvious in view of the unprecedented benefits it offers [6–8]. One of the most striking advantages of nanoparticles (NPs) is a net increment in the surface area-to-volume ratio due to their reduced size which results in a higher product yield and also greater selectivity.

---

*Corresponding author.

Indeed, it is undoubtedly the size-related property that has offered uncountable opportunities for surprising discoveries [9–13]. Industries have been deriving the utmost benefits with the design of tunable NP-based catalytic systems that are being exploited for different organic reactions [14, 15]. Today, undeniably nanoparticle research has emerged as a fascinating branch of science.

Since the pioneering report by Stöber, silica nanoparticles (SiO_2 NPs or SNPs) also known as the nano form (<100 nm) of silicon dioxide have emerged as an attractive support material for organic–inorganic hybrid heterogeneous catalysts [16]. In fact, the chemical compound silicon dioxide or silica (SiO_2), which can be present in either crystalline or amorphous forms, is one of the most abundant components of the earth's crust found in nature. It has been known since antiquity. Silicon was first isolated and described as an element in the year 1824 by a Swedish chemist Jons Jacob Berzelius.

Jons Jacob Berzelius — The man behind the discovery of the element Si

It took almost 30 years to produce the crystalline form of SiO_2 and this was accomplished using an electrolysis technique. Studies have revealed that SiO_2 frequently exists in the form of a three-dimensional polytetrahydral network structure (Figure 2.1), built up by packing [SiO_4] units, and a bulk structure that terminates at the surface in two different ways, that is, with oxygen on

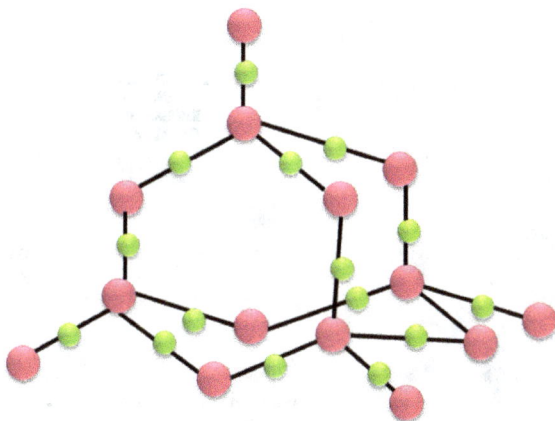

Figure 2.1. Giant structure extending in all the dimensions.

the surface through siloxane groups (μSi–O–Siμ) or silanol groups (μSi–OH). The crystalline form of SiO$_2$ has a structural similarity with diamond [17].

Over the years, SiO$_2$ NPs have generated a considerable amount of attention owing to their unique and adjustable physicochemical properties, such as high surface area, nanometer size and rigid framework, with excellent chemical, thermal and mechanical stabilities [18–21]. When SiO$_2$ NPs are employed as solid supports for the fabrication of organic–inorganic hybrid nanocatalysts, they exhibit several outstanding features, such as high activity and selectivity, excellent stability, efficient recovery and recyclability. An additional feature that multiplies their use manifold as an exceptional support material is the possibility to functionalize both the exterior and interior pore surfaces with several organic moieties. To date, there have been many reports wherein SNPs have been used to develop highly active and selective, recoverable catalysts for various organic transformations. For example, Sharma *et al.* have developed a silica-based organic–inorganic hybrid copper catalyst for oxidative amidation of methyl ketones [22]. The nanostructured copper catalyst exhibited high efficacy with excellent functional group tolerance and allowed the generation of a wide range of amidation products under mild reaction

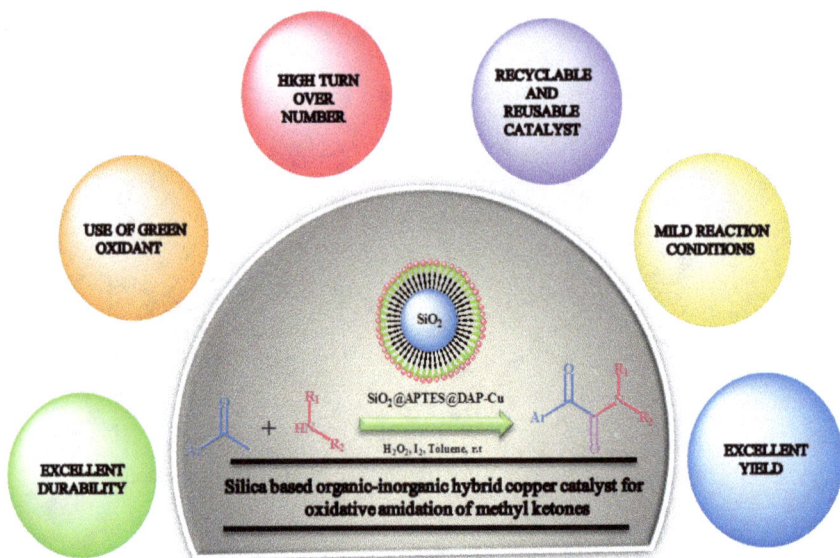

Figure 2.2. Silica-based organic–inorganic hybrid copper catalyst for oxidative amidation of methyl ketones.

conditions. In addition, the catalyst exhibits remarkable durability as well as good recyclability for several runs with a high consistency in its catalytic activity (Figure 2.2).

2.2 Surface Chemistry of Silica

Primarily, there are three types of silanol groups, namely (Figure 2.3, [23]),

  (i) vicinal (hydrogen-bonded silanols),
 (ii) geminal (two hydroxyl groups attached to the same silicon atom),
(iii) isolated (no hydrogen bonds possible).

It is the presence of large concentration of surface silanol (Si–OH) groups that allows the straightforward functionalization of the exterior and interior pore surfaces of the SiO_2 NPs with suitable linking agents. Literature reports document that silanol groups can

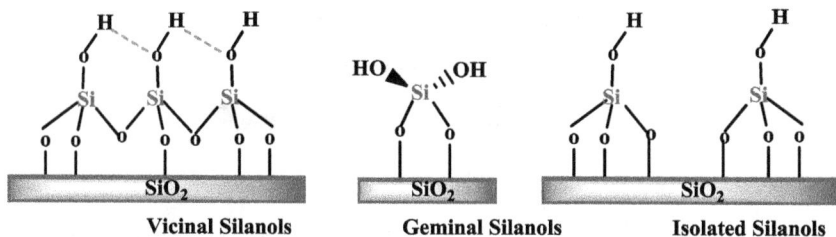

Figure 2.3. Various types of silanol groups present on the surface of silica nanospheres (SNSs).

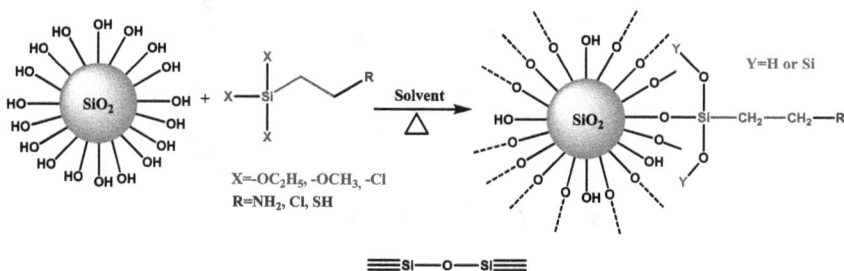

Figure 2.4. Surface modification of SNPs.

be readily functionalized using different chemical methodologies [24, 25]. Among a wide variety of surface modification techniques, the reaction between silanol groups and silane reagents such as (3-aminopropyl)triethoxysilane (APTES) results in the effective functionalization of the silica surface (Figure 2.4) [21]. Undeniably, it is their rich surface chemistry and a tailorable morphology that multiplies their use manifold as a support material. The detailed studies related to surface modification shall be taken up in Chapter 4.

2.3 Classification of SNPs

SNPs can be classified as mesoporous and non-porous (solid) NPs, both of which bear the amorphous silica structure. While mesoporous silica NPs (MSNs) are characterized by the mesopores (2–50 nm pore size), non-porous ones as the name suggests are devoid of any pores. MSNs were independently synthesized in 1990 by researchers in Japan. Six years later, silica nanoparticles with much larger

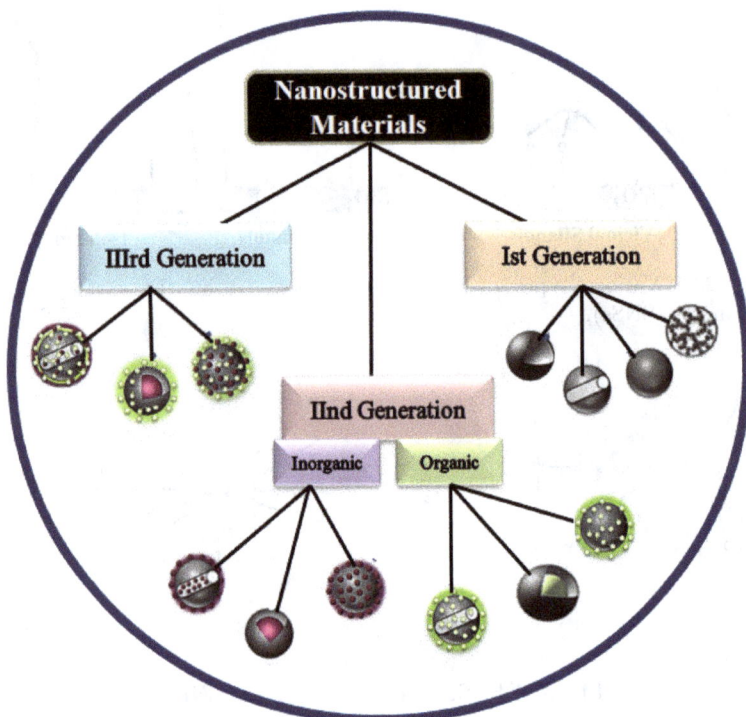

Figure 2.5.   Classification of NSMs.

(4.6–30 nm) pores were produced at the University of California, Santa Barbara [26].

Recently, on the basis of important parameters such as composition, extent and location of surface treating agents, a classification system (using nanostructured silica materials (NSMs) as an example) has been established (Figure 2.5) [27]. This system enables systematic classification of a broad spectrum of silica materials and is also applicable to even metals and oxide-based manufactured nanomaterials (MNMs) in general.

2.3.1   First-generation nanostructured silica materials (1G-NSMs)

The very first classification, i.e. 1G-NSMs, includes only bare SNPs where either the surface or the internal structure of amorphous silica consists of end groups, such as silanols or siloxanes. Silica aerogels, fumed silica and porous SNPs are some of the examples of the 1G-NSMs.

2.3.2 Second-generation nanostructured silica materials (2G-NSMs)

The 2G-NSMs include composites of nanostructured silica that further comprise the following:

(1) one or more organics called as *nanocomposite of silica-organic (2GO)*,
(2) one or more inorganic compounds as their secondary phase called as *nanocomposite of silica-inorganic (2GI)*.

2.3.3 Third-generation nanostructured silica materials (3G-NSMs)

This category includes one or more organics and one or more inorganic compounds as their counterparts. Few examples of the 3G-NSMs are silica-organic/inorganic nanocomposites, inorganic/organic-silica nanocomposites and organic/inorganic-silica-inorganic/organic nanocomposites.

2.4 Synthesis of SNPs

Perceiving the innumerable benefits offered by the absolutely riveting SNPs, remarkable efforts have been devoted by the scientific community toward the investigation of novel processing methodologies that are utilized to synthesize them. Judicious selection of different experimental parameters can result in better control of the size, shape, porosity and significant improvement in the physicochemical properties.

The following techniques are available for preparing the SiO_2 NPs:

(i) Stöber's method (also known as the sol–gel approach),
(ii) microemulsion,
(iii) hydrothermal synthesis,

(iv) ball milling,

(v) flame-spray pyrolysis,

(vi) chemical vapor deposition.

Among these, preparation by sol–gel has attracted the synthetic pursuit of researchers as it is straightforward, scalable and controllable.

2.4.1 Stöber's method (sol–gel approach)

Long back in the year 1956, Kolbe observed the formation of silica particles when tetra-ethylorthosilicate (TEOS) was made to react with water in the presence of a base [28]. Thereafter, Stöber and Fink in 1968 reported the development of a system of chemical reactions that controlled the growth of silica particles [29]. Today, it remains one of the most widely employed methodologies to prepare SNPs and is also referred to as the Stöber's approach/sol–gel approach. As shown in Figure 2.6, the sol–gel technique involves the use of alkoxysilanes such as TEOS or inorganic salts such as sodium silicate (Na_2SiO_3) as silica precursors that are hydrolyzed to monomeric $Si(OH)_4$ in the presence of polar solvents such as ethanol [30].

The $Si(OH)_4$ molecules then condense to produce silica oligomers *via* siloxane groups that finally end up in an amorphous silica network (the steps have been elucidated in the flowchart depicted in Figure 2.7).

Both these steps, hydrolysis and condensation, occur simultaneously and the use of a basic catalyst allows the entire process to be kinetically faster. Ammonia functions as a catalyst in the

Hydrolysis:

$$Si(OC_2H_5)_4 + 4H_2O \longrightarrow Si(OH)_4 + 4C_2H_5OH$$

(TEOS)   (Silicon tetra hydroxide)

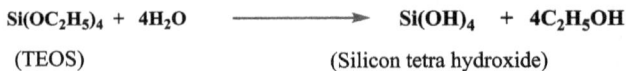

Polycondensation:

$$Si(OH)_4 \longrightarrow \text{Silica Nanoparticles} + 2H_2O$$

Figure 2.6. Sol–gel synthesis of SNPs.

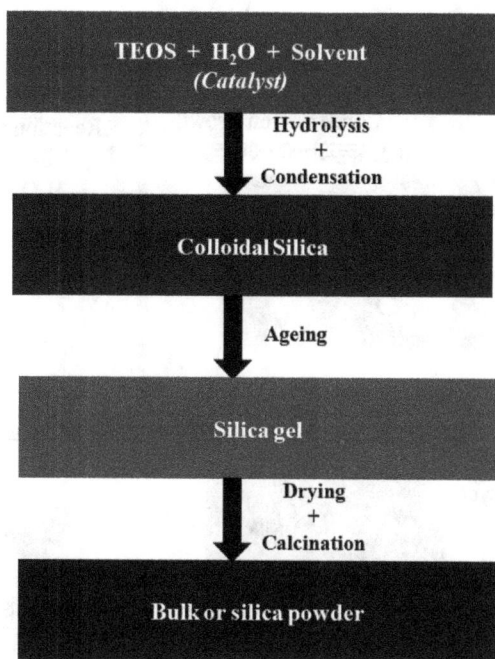

Figure 2.7. Flowchart of a typical sol–gel process.

hydrolysis of TEOS in ethanol and generates the monodispersed solid SiO_2 NPs (size ranging from 50 nm to 2 μm). Although alcohols like methanol and propanol can also be utilized as solvent for this synthesis, the reaction in ethanol generates the finest particles that are generally monodisperse, spherical in shape and hydrophilic in nature. This methodology allows the synthesis of both nano-sized non-porous as well as MSNs in low temperature and under milder reaction conditions [31]. The mechanism of the formation of the SNPs involves two stages, namely, nucleation and growth. To describe the growth mechanism of silica, two models have been proposed, both of which lead to the generation of either spherical or gel network depending on the reaction conditions (Figure 2.8).

• **Monomer Addition:** According to this model, growth primarily occurs on the surface of the primary particles *via* the addition of hydrolyzed monomers usually after an initial burst of nucleation.

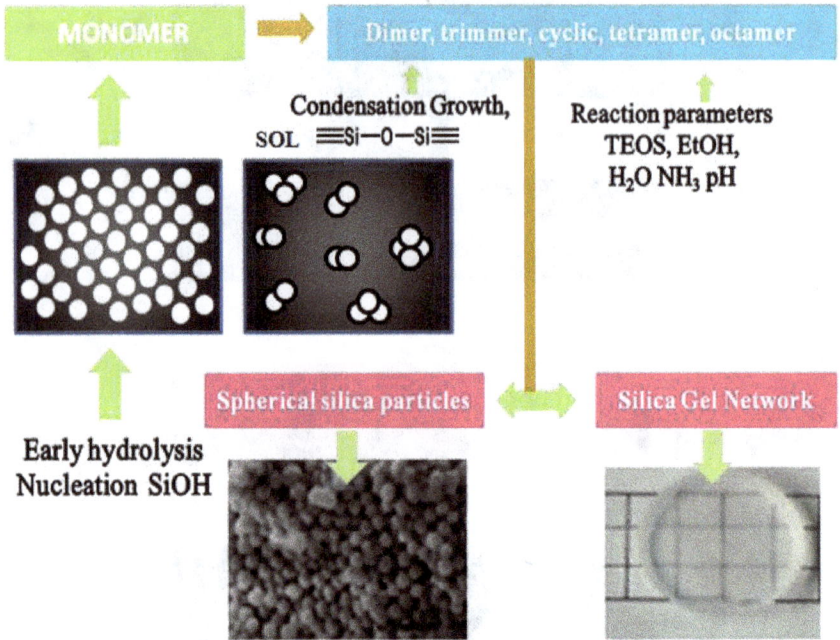

Figure 2.8. Description of sol–gel process leading to the formation of silica (adapted from Ref. [31]).

- **Controlled Aggregation:** The aggregation model elaborates that the nucleation occurs continuously throughout the reaction, and the resulting nuclei (primary particles) will aggregate together to form dimers, trimers and larger particles (secondary particles).

In fact, many contemporary research studies on the synthesis of nanosilica particles have indeed evolved from Stöber's method [32–38]. Many researchers have attempted to study the effect of reaction variables on the growth of the nanoparticles. Recent works have shown that with the increase in the concentration of ammonia, an increase in the particle size can be achieved (Figure 2.9) [31].

For the preparation of uniform-sized SNPs, recently an amino acid-catalyzed methodology has been reported, which involves the hydrolysis and condensation of TEOS in an aqueous system under weak basic conditions [39]. This methodology offers a significant

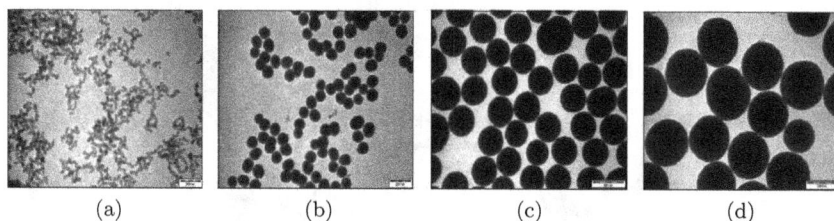

Figure 2.9. Different sizes of silica obtained by controlling reaction parameters: (a) ~21 nm, (b) ~131 nm, (c) ~369 nm and (d) ~565 nm (Adapted from Ref. [31]).

advantage over the other synthetic approaches as it employs water as the solvent that has been demarcated as a green solvent by various international bodies, such as ACS Green Chemistry Institute®. Thus, one finds that this methodology avoids the use of harmful/toxic organic solvents. Apart from the conventional base-promoted synthesis of SNPs, acid-catalyzed hydrolysis has also been employed by Sadek and co-workers [40]. This research group carried out the hydrolysis of TEOS using hydrochloric acid and could successfully synthesize SNPs. Recently, Lovingood and co-workers have developed a surfactant-free, controlled, microwave-assisted methodology for the synthesis of SNPs, having diameters ranging from 30 to 250 nm [41].

Literature reports have shown that the sol–gel methodology has also been utilized for the synthesis of MSNs, which is a wondrous class of NPs that have attracted substantial interest for their use as catalyst-immobilization matrices as they possess controlled particle size, porosity, morphology and high chemical stability. Tang *et al.* have prepared sub-micrometer MCM-41 spherical particles by modifying the Stöber synthetic compositions *via* the addition of a cationic surfactant to the reaction mixture [42]. Similarly, there are other reports on the fabrication of uniform MSNs with different pore sizes and mesostructures, wherein mixtures of alcohol, water and ammonia have been utilized with different template systems.

Recently, a simple and facile one-pot sol–gel method has been reported by Sun and research group for the fabrication of hollow mesoporous silica particles (Figure 2.10) [43]. The synthesis was

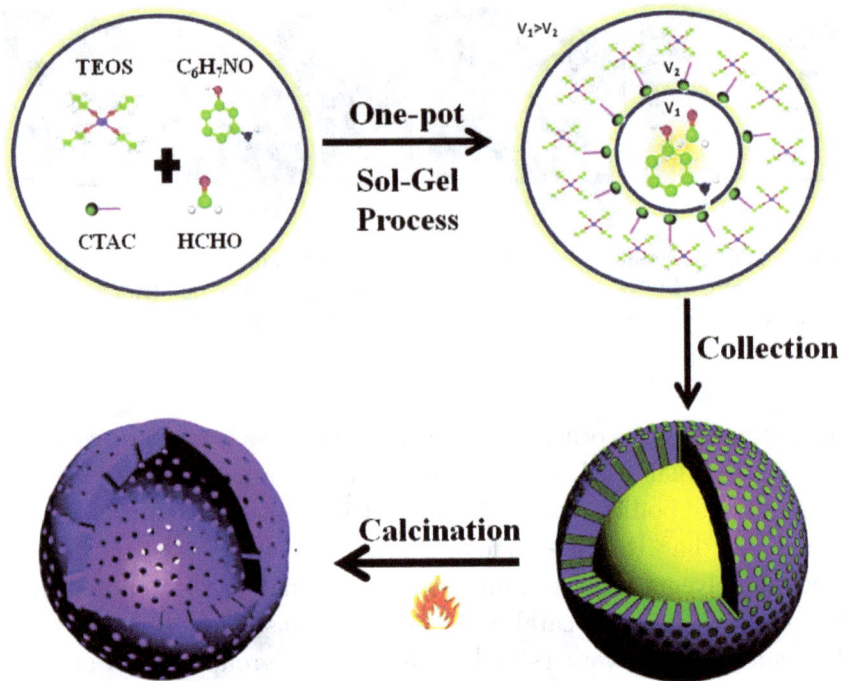

Figure 2.10. Preparation strategy of the hollow mesoporous silica spheres by using a facile one-pot sol–gel process (adapted from Ref. [43]).

carried out by reacting TEOS, 3-aminophenol, formaldehyde, hexadecyl trimethylammonium chloride (CTAC) and ammonia in a mixture of ethanol and water. They observed that this technique allowed the control in particle size and the shell thickness *via* moderately tuning some experimental parameters (for instance, simply by changing the amount of silica precursors, as shown in Figure 2.11).

2.4.2 Microemulsion synthesis (MES)

The word "microemulsion" was originally proposed by Schulman and co-workers in the year 1959 [44]. Since the emergence of the concept of "microemulsions", the MES technique has become one of the most versatile preparative methods that has found a wide range of applications right from oil recovery to synthesis of nanoparticles as it

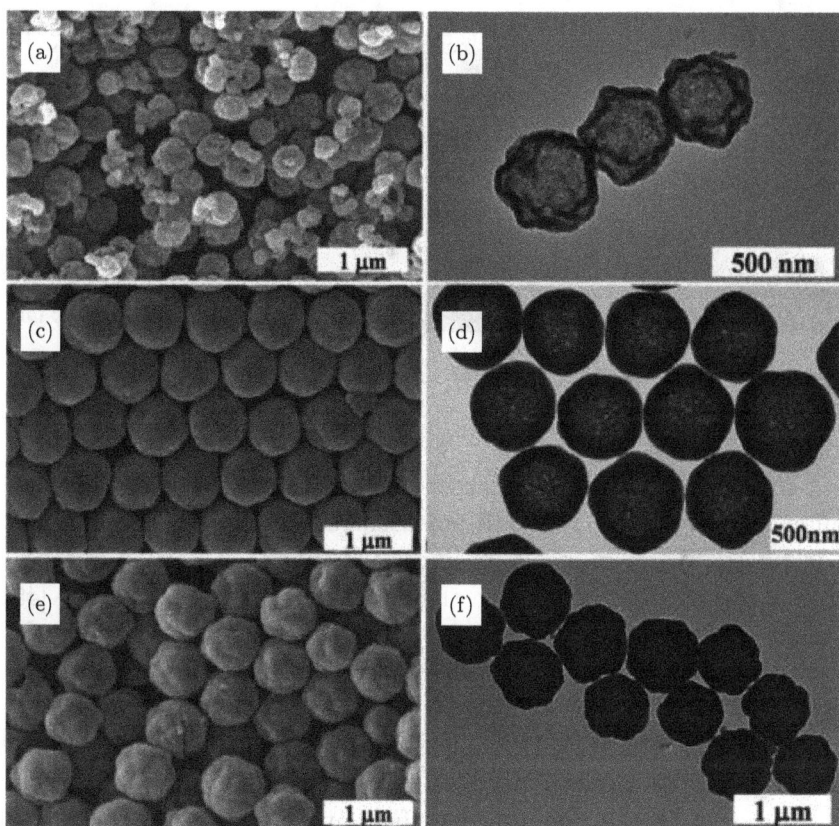

Figure 2.11. FESEM images (a), (c) and (e) and TEM images (b), (d) and (f) of hollow silica particles prepared at different concentrations of TEOS: (a) and (b) 0.36 ml, (c) and (d) 1.08 ml, (e) and (f) 1.44 ml (adapted from Ref. [43]).

enables to control particle properties, such as mechanisms of particle size control, geometry, morphology, homogeneity and surface area. Microemulsions are defined as thermodynamically stable dispersion of immiscible fluids containing at least three components, namely, a polar phase (usually water), a non-polar phase (usually oil) and a surfactant. In comparison to the usual emulsions, microemulsions can be formed by simply mixing the desired components and they do not need high shear conditions that are generally involved in the formation of ordinary emulsions.

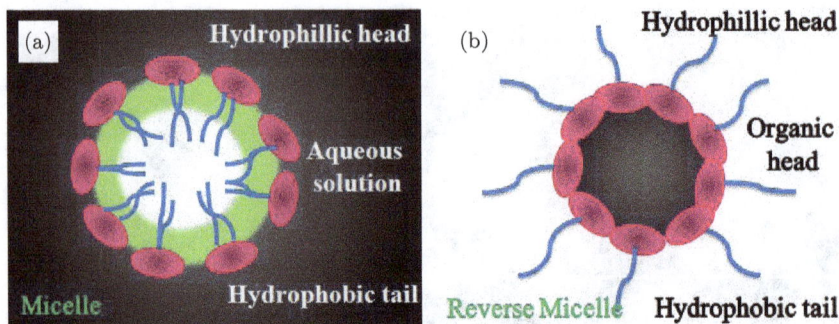

Figure 2.12. Pictorial description of (a) micelle (oil-in-water) and (b) reverse micelle (water-in-oil).

Broadly speaking, in microemulsions, the surfactant molecules align themselves to generate spherical aggregates in the continuous phase. Water-in-oil (w/o) microemulsions, or reverse micelles, have generally been used for the synthesis of ultrafine SNPs [45] (Figure 2.12).

SNPs start growing inside the water droplet of a w/o microemulsion and this process is catalyzed *via* the base-assisted hydrolysis of silicon alkoxide (Si-OR'). However, the major drawbacks associated with this approach include exorbitant cost and complexities associated with the removal of surfactants from the final products. Arriagada and Osseo-Asare reported the synthesis of SNPs through the controlled hydrolysis of tetraethoxysilane in a non-ionic surfactant/ammonium hydroxide/cyclohexane reverse micellar system [46].

Likewise, Finnie *et al.* have reported the preparation of SNPs for controlled-release applications under both acidic and basic conditions *via* the reaction of tetramethylorthosilicate (TMOS) inside the water droplets of a w/o microemulsion [47]. The *in situ* Fourier-transform infrared spectroscopy (FTIR) measurements were carried out which revealed that when TMOS is added to the microemulsion, silica is formed as TMOS that is preferentially located in the oil phase, which then diffuses into the water droplets. In the hydrophilic phase, hydrolysis takes place rapidly due to high local concentration of water. Under basic conditions, the small-angle X-ray scattering (SAXS) analysis

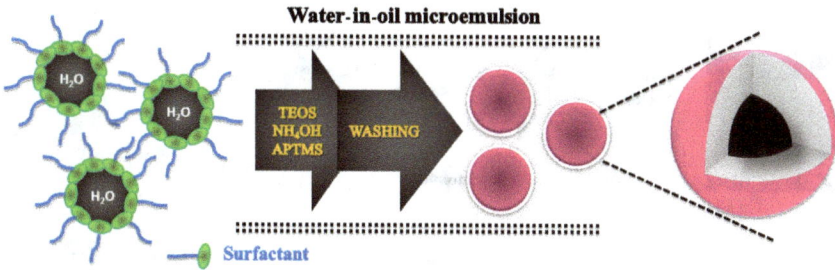

Figure 2.13. Fabrication of HSNs using w/o reverse microemulsions.

revealed that approximately 11 nm non-porous spheres are formed which are stabilized by a water/surfactant layer on the particle surface; whereas under acidic conditions, highly uniform ~5-nm porous spheres get generated, which appear to be retained within the water droplets.

Recently, hollow silica nanospheres (HSNs) have also been synthesized using w/o reverse microemulsions primarily containing the polymeric surfactant "polyoxyethylene" isooctylphenyl ether (Igepal CA-520), ammonia and water in a continuous oil phase (alkanes) [48]. SNPs were formed by the nucleation of ammonium-catalyzed silica oligomers from silica precursors — "tetraethylorthosilicate (TEOS)" and "nanoporous aminopropyltrimethoxy silane (APTMS)" in the reverse microemulsion system (Figure 2.13).

Even for the purpose of silica coating of gold nanoparticles, the reverse microemulsion method (water in cyclohexane reverse microemulsion) has been reported, but no silane coupling agent or a polymeric material was used as the surface primer for this purpose [49]. It was found that the prepared silica-coated gold nanoparticles were overall 24 nm in size.

Apart from this, core–shell Au@SiO$_2$ spherical nanoparticles comprising multiple gold nanodots (each with a maximum diameter of 5 nm) and silica shell have also been fabricated by Yoo and co-workers using a reverse w/o microemulsion-based synthetic approach (Figure 2.14) [50]. The microemulsion primarily consisted of Brij-35 (surfactant) cyclohexane and n-hexanol which was prepared under ultrasonic irradiation. Thereafter, the gold precursor (aq. HAuCl$_4$),

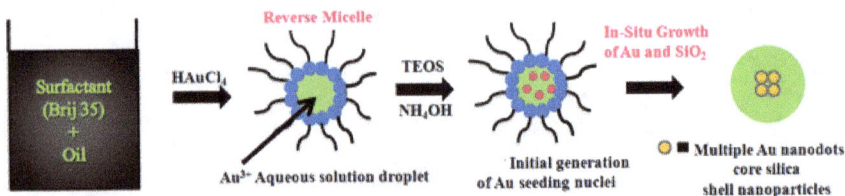

Figure 2.14.   Synthesis of multi-AuSiO$_2$ NPs in a reverse microemulsion.

silica precursor (TEOS) and a base (NH$_4$OH) were added sequentially to the reverse microemulsion and the reaction contents were subjected to further stirring at room temperature. On completion of the reaction, the microemulsion system was destabilized by adding acetone and the as-synthesized multi-AuSiO$_2$ NPs were purified by repeated washing. The entire synthetic process described above enabled the *in situ* growth of Au nanoparticles which got suitably encapsulated within the silica layer (Figure 2.15).

2.4.3   Hydrothermal synthesis

Among a wide diversity of synthetic approaches, hydrothermal synthesis has gained tremendous interest of researchers from multidisciplinary fields over the past few years. As the name suggests, "hydro" means water and "thermal" means heat, this method relies on the use of high-temperature and high-pressure (high TP) water conditions for the formation of crystals and therefore is carried out in autoclaves (thick-walled sealed steel cylinders) that can withstand these high TP conditions for a long time. It was the great British geologist, Sir Roderick Murchison, who first used this word for describing the action of water at elevated temperature and pressure in bringing about changes in the earth's crust that resulted in the formation of different rocks and minerals [51]. The underlying principle is that most of the inorganic substances dissolve in water at an elevated temperature and pressure which eventually leads to the re-crystallization of a substance to be synthesized from the dissolved material.

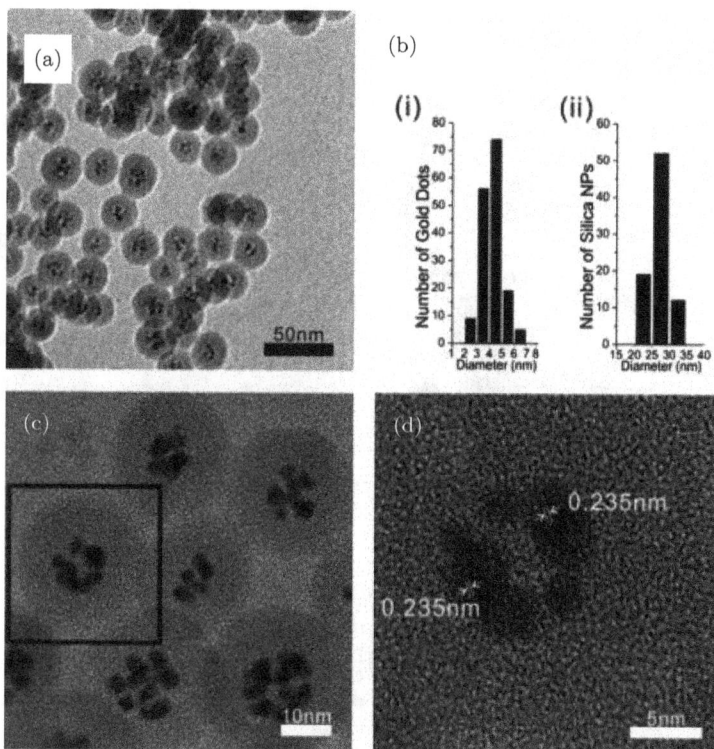

Figure 2.15. (a) Transmission electron microscopy (TEM) image of the multi-Au@SiO$_2$ NPs that were synthesized using 0.03 mL of 1 M HAuCl$_4$ (aq.) (0.03 mmol); (b) size (diameter) distribution of the corresponding (i) Au nanodots and (ii) SNPs; (c) and (d) high-resolution TEM images of the multi-Au@SiO$_2$ NPs ((d) is a highly magnified image of the particle enclosed within the box shown in (c)) (adapted from Ref. [50]).

2.4.3.1 *Why water?*

- The primary role of water is to cause the transformation of the precursor materials, and this can be done because of excessive vapor pressure and its structure being different from that at room temperature [52]. Thus, it can act as a catalyst for the formation of the desired materials by tuning the temperature and pressure.
- Advantages of using water as the solvent include the following:
  - (i) Water is environmentally friendly and cheap in comparison to the other solvents.

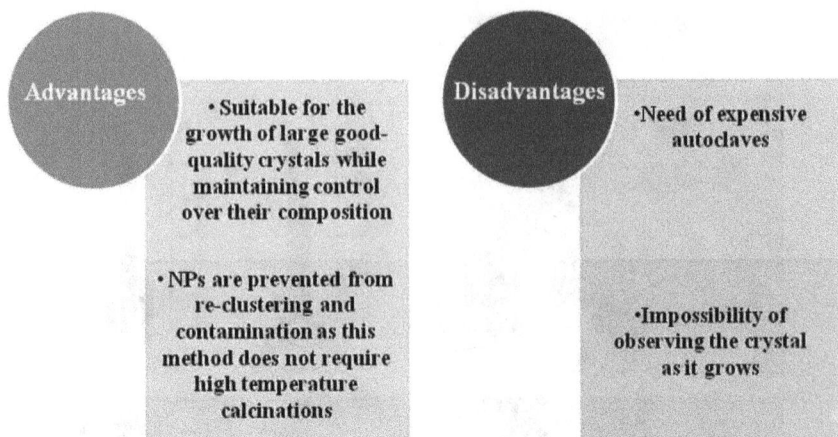

Figure 2.16. Advantages and disadvantages of the hydrothermal synthetic approach.

(ii) It is non-toxic, non-flammable, non-carcinogenic, non-mutagenic and thermodynamically stable.
(iii) Since water is very volatile, it can be removed from the product very easily.

The key parameters that define both the process kinetics and properties of resultant products prepared through this method include the following:

• initial pH of the medium,
• duration and temperature of synthesis,
• pressure in the system.

Advantages and disadvantages of the hydrothermal synthesis technique have been summarized in Figure 2.16.

Although this method has been extensively employed for the synthesis of different types of NPs, it has been found to be particularly effective for the synthesis and post-synthetic treatment of MSNs. MSNs were first discovered in the 1990s by Mobil scientists and since then they have attracted great interest. It was found that using the hydrothermal synthesis method, MSNs can be generated with higher hydrothermal stability, improved mesoscopic regularity and

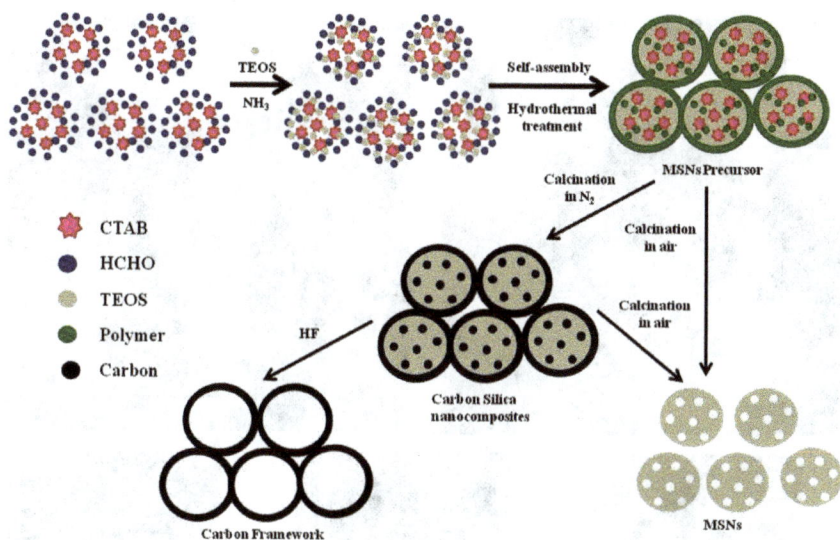

Figure 2.17. Schematic illustration of the formation process of MSNs.

high pore size. Recently, Gu *et al.* have fabricated MSNs through a facile one-pot suppressed growth hydrothermal methodology in which water/formaldehyde solution system was utilized as the solvent, cetyltrimethylammonium bromide (CTAB) as the template, aqueous ammonia as the basic catalyst and TEOS as the silica source (Figure 2.17) [53]. It was found that formaldehyde worked as a suppressant thereby controlling the growth of SNPs.

Apart from MSNs, hollow silica spheres (HSSs) have also emerged as excellent support material for the synthesis of supported heterogeneous catalytic systems. Recently, Wang and co-workers fabricated HSSs of different sizes using a self-templating method in acidic aqueous media [54]. They observed that the reaction time and concentration of silica played a vital role in controlling the hollow interior space. Additionally, salts such as NaCl and Na_2SO_4 not only resulted in faster hollowing but also helped in the production of well-defined shell structures (Figure 2.18). Figure 2.18(c) depicts the TEM images of silica spheres etched in the HCl/Na_2SO_4 system; this figure reflects the formation of an intermediate with rattle-type morphology. The relatively small spheres could evolve into hollow

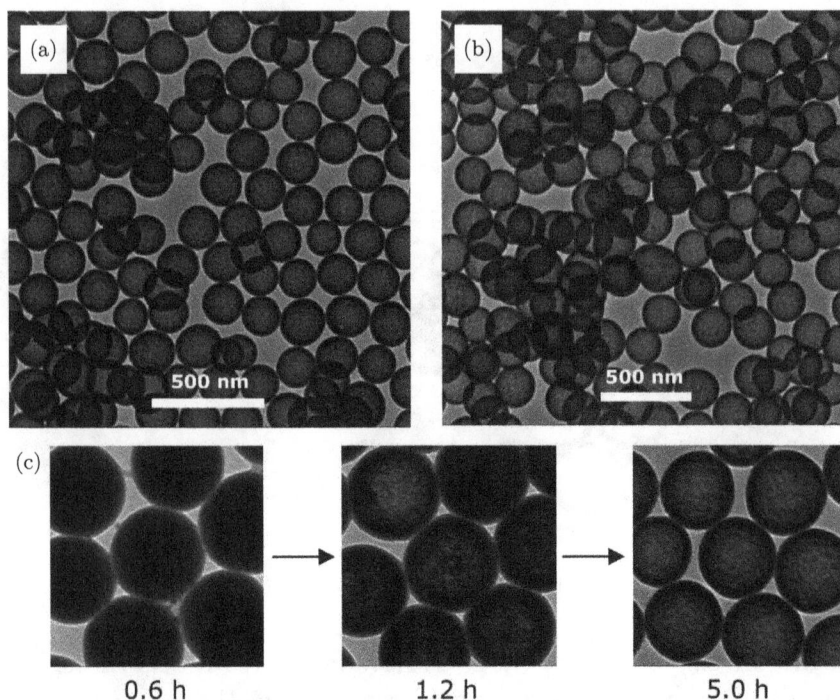

Figure 2.18. TEM images of the HSSs prepared in (a) HCl/Na$_2$SO$_4$ and (b) HCl/NaCl; (c) TEM images of silica spheres etched in HCl/Na$_2$SO$_4$ system at different reaction times (adapted from Ref. [54]).

spheres *via* control in the reaction time; however, in the case of the larger spheres of size 540 nm, not much change in the morphology could be observed. Besides, the pore structure of HSSs could be controlled by tuning the acidity of the silica dispersion. This was for the first time that an acidic medium was involved in the generation of HSS.

Very recently, Qisti and Indrasti reported that bagasse ash which is a solid waste of burned bagasses has a high silica content and therefore can be used for the production of nanosilica [55]. The hydrothermal synthesis pathway was undertaken for synthesizing the SNPs as it offered innumerable benefits, such as facile preparation, mild reaction temperature, uniform dispersion of the doping metal ions, stoichiometry control and good chemical homogeneity.

2.4.4 Green synthesis

The intense concern over the hazardous pollutants and global energy crisis during recent times has surged the development of environmentally benign materials based on natural and renewable raw materials. Thus, now there is an increasing emphasis on the biogenic (plant-mediated especially) synthesis of NPs which has emerged as a Green Chemistry approach interconnecting nanotechnology with biology [56, 57]. The approaches involving green synthesis of NPs generally follow the following goals:

- synthesis at ambient temperatures,
- neutral pH conditions,
- low costs,
- use of environmentally friendly reagents.

Owing to the high silica content, low cost and widespread availability, Tekinay *et al.* have used sugarbeet bagasse as the starting material for the generation of SNPs [58]. They employed laser ablation technology and successfully synthesized the target SiO_2 NPs within a single pot in a single step. As demonstrated in Figure 2.19, the following two different methodologies were employed for the synthesis of SiO_2 NPs:

(1) calcination of sugarbeet bagasse and the subsequent treatment of sugarbeet bagasse ash with NaOH;
(2) synthesis using laser ablation.

Experiments done using these methodologies revealed that the synthesis of SiO_2 NPs using the laser ablation technique not only resulted in the generation of NPs of smaller size but also had a positive effect on the growth of algae (*Calluna vulgaris*) which suggested that they had no harmful effects on the aquatic environment.

Apart from sugarbeet, sugarcane waste ash [59] and rice husk [60] have also recently come into limelight as raw materials containing relatively high silica content for the preparation of silica-based NPs (Figure 2.20). The use of such raw materials shows huge prospects of recovering wealth out of it.

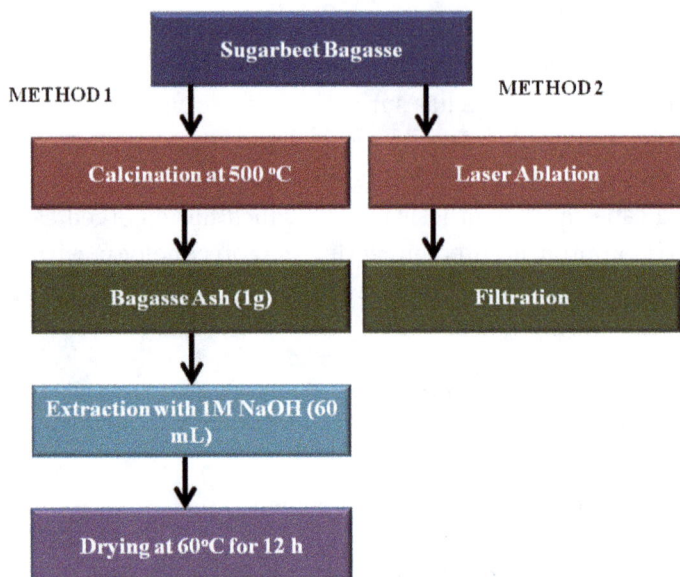

Figure 2.19. Flowchart describing the production of SNPs from sugarbeet bagasse.

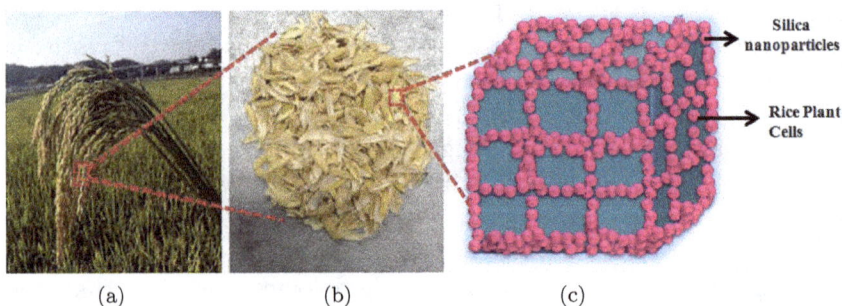

Figure 2.20. (a) Rice grown on farm, (b) rice hulls and (c) structural framework of rice hull.

Vijayalakshmi and co-workers thought of utilizing this natural resource, i.e. rice husk for obtaining SiO_2 NPs [61]. For doing so, first, they sintered rice husk at 900°C for several hours and then treated it with 1M NaOH to form sodium silicate. The resulting gel was then taken and reacted with 6M H_2SO_4 for the precipitation of SNPs.

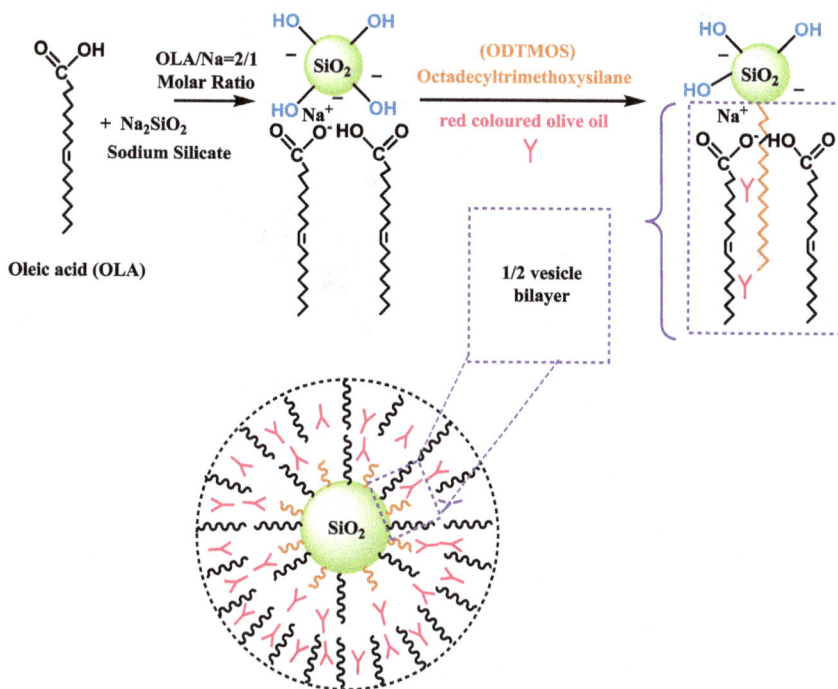

Figure 2.21. Schematic representation of the synthesis of MSNs stabilized with an oleic acid/sodium oleate complex starting from sodium silicate and octadecyltrimethoxysilane.

Ianchis *et al.* reported for the first time the possibility of synthesizing MSNs from natural raw materials: "sodium silicate" from sand and "oleic acid" (OLA) from saponified vegetable oils (Figure 2.21) [62]. The moment these two components interact, sodium neutralizes a fraction of OLA, forming its alkaline salt/soap (OLANa). After this process of neutralization, the pH falls down, allowing the SNPs to be generated. It was found that the average diameter of the ensuing particles increased significantly at a molar ratio of OLA/OLANa >2/1.

2.4.5 Microwave-assisted synthesis of SNPs

The growing awareness and recognition for alternative eco-friendly and economical protocols have paved the pathway for the

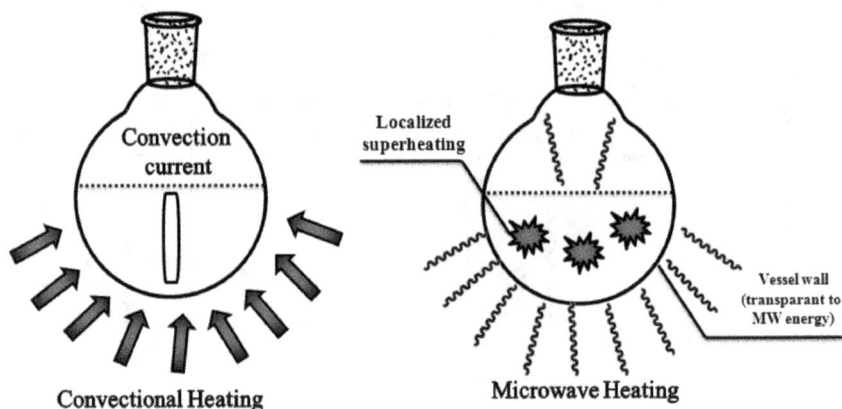

Figure 2.22. Comparison of the conductive and microwave heating methodologies.

development of microwave-assisted reactions for the synthesis of valuable products and entities. In fact, such types of reactions have emerged as one of the most powerful and sustainable tools in synthetic chemistry. In comparison to the conventional heating methodologies, the microwave reactors enable faster heating of reaction mixtures to high pressures and temperatures and thus leading to reduction of reaction times, improved yields, modifications of selectivities, increased product purities, simplification of work-up procedures and efficient atomic utilization (Figure 2.22) [63].

Recently, Lovingood and co-workers [41] demonstrated a surfactant-free, acid-catalyzed microwave-assisted synthetic route for obtaining SNPs with controlled size and morphology. TMOS was chosen as the silica precursor for synthesizing the SiO_2 NPs [41]. The first step of this microwave reaction involved the hydrolysis of TMOS in the presence of 1 mM HCl which lead to the generation of silicic acid. The as-obtained silicic acid was thereafter diluted using acetone, and different mixtures with volumetric ratios of silicic acid to acetone (1:66, 1:50, 1:40, 1:33, 1:28, 1:25, 1:22, 1:20) were prepared. Five-milliliter aliquots of each of these mixtures were placed in a 10-mL CEM vial containing a stir bar and snap cap. Finally, under microwave irradiation conditions, the SiO_2 NPs were synthesized. The experimental results revealed that a proper selection of solvent, silicic acid precursor, catalyst and microwave irradiation time played a crucial role in controlling the size of the NPs. Strikingly, this methodology showed a large-scale utility in the production of SiO_2 NPs for various industrial applications.

Santiago and research group have also fabricated hydrophobic SiO_2 NPs using a two-step microwave-assisted sol–gel method [64]. The first step involved the generation of SiO_2 NPs of different sizes from TEOS, while the second step involved the hydrophobization of the NPs using hexadecyl trimethoxysilane (HDTMOS). The utilization of microwave irradiation technology resulted in high conversion degrees in a relatively very short reaction time.

Similar methodology has been utilized by Mily *et al.* [65] also. The SiO_2 NPs have been synthesized using the sol–gel microwave-assisted methodology in alcoholic solution in the presence of an ammonium catalyst. Observation of experimental results showed that the particle size and conversion could be controlled by simply varying the concentration of water and ammonia. Also, they found that in contrast to the traditional heating methods, this microwave irradiation methodology resulted in a higher reaction rate and narrower particle size distribution.

The microwave irradiation technology has also been used for the synthesis of MSNs under different heating powers (100, 300 and 450 W) [66]. For obtaining the MSNs, first, CTAB, ethylene glycol (EG) and NH_4OH were dissolved in water and, after vigorous heating, TEOS was added to the reaction mixture which led to the formation of a white micelle solution. This solution was stirred continually for about 2 h and transferred to a beaker which was then kept in the microwave oven for varying time intervals (8, 4 and 1 hours under heating power settings of 100, 300 and 450 W, respectively).

2.4.6 Synthesis using continuous-flow reactors

Although both the academic as well as industrial sectors have been relying on the batch operations for performing synthetic chemistry since decades, one problem with these conventional reactor technologies that has been puzzling the researchers continually is the failure to

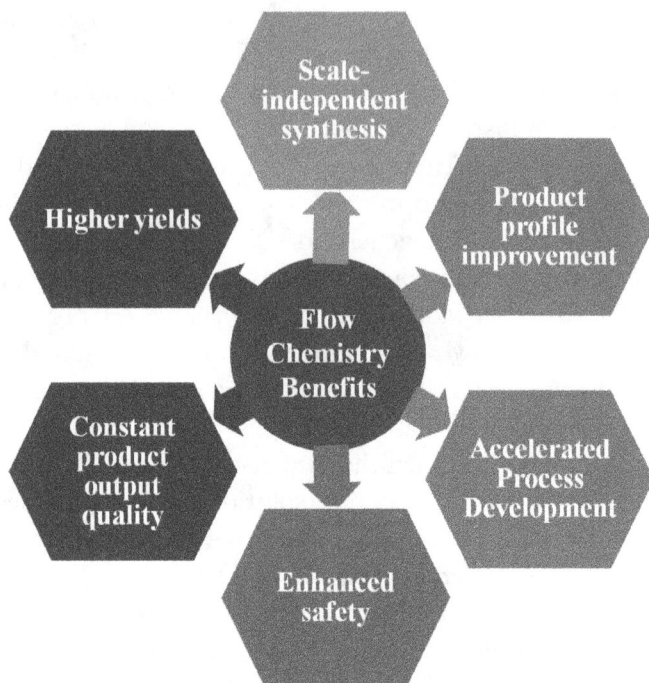

Figure 2.23.   Advantages of flow synthesis.

reproduce and scale-up these reactions. This problem is particularly relevant for exothermic processes. However, application of microwave technology and continuous-flow chemistry has now been widely accepted as a powerful solution for overcoming these problems *via* the large-volume production demonstrated through the replication of unit processes. In contrast to the batch method, flow chemistry involves the continuous pumping of the reagent streams into a flow reactor where they are allowed to mix and subsequently react [67, 68]. The product formed during this step instantly leaves the reactor as a continuous stream; thus, the synthesis scale is governed only by the flow rate and the operation time. As a result of the large-scale benefits offered by it, the continuous-flow processing has emerged as a key enabling technology, transforming the way chemistry is conducted and expanding our horizon of synthetic capabilities (Figure 2.23).

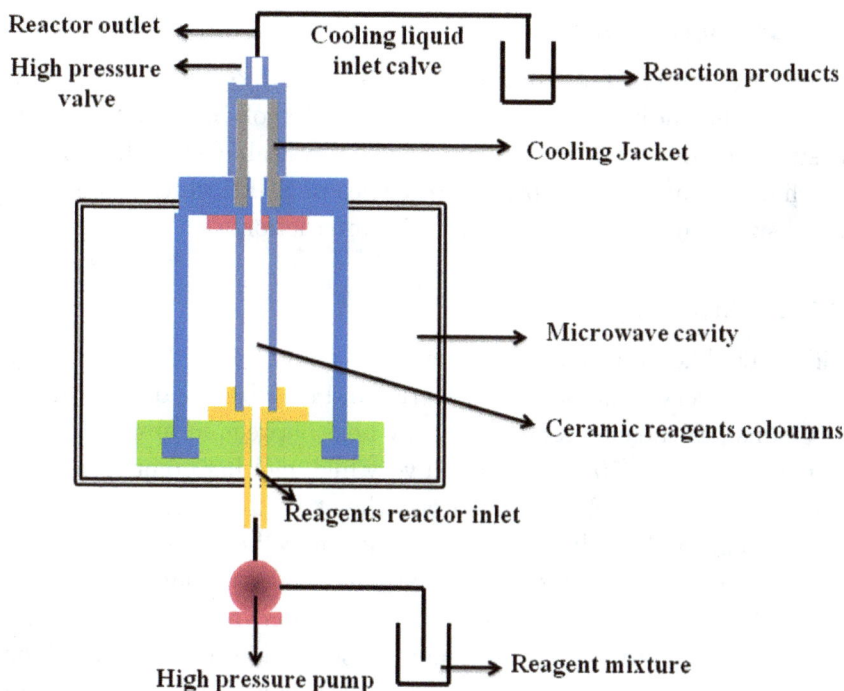

Figure 2.24. Schematic diagram of the continuous-flow microwave reactor for the synthesis of SNPs.

Very recently, continuous-flow synthesis has also been effectively exploited in the synthesis of different structured nanomaterials. A continuous-flow microwave reactor was successfully used by Corradi and other co-workers for preparing monodisperse colloidal spherical nanoparticles *via* the hydrothermal synthesis approach that involved the hydrolysis and condensation of TEOS (Figure 2.24) [69]. In order to obtain the unagglomerated SiO_2 NPs, the flow rate was regulated between 43 and 101 mL/min. When compared to the batch method, it was found that this methodology could generate particles of 50 nm in size which did not undergo agglomeration as the residence time was decreased by increasing the flow rate to 101 mL/min. In addition to this, continuous-flow microwave synthesis also exhibited a shorter reaction time.

2.5 Properties of SNPs

It is a well-known fact that at nanoscale, the properties differ from that of the individual atoms and molecules of bulk matter. The ensuing sections shall throw light on the physical, chemical, thermal, mechano-chemical and optical properties of SNPs which depend on their size, composition and interfacial interaction.

2.5.1 Physical properties

Silica acts like an insulator as it exhibits poor electrical conductivity owing to a very large band gap energy and the absence of delocalized electrons (all the electrons are tightly held between atoms and hence not free to move) [70]. It also has a very high melting point (generally about 1700°C) which depends upon the structural form in which it exists. The high melting point of silica may be attributed to the presence of very strong silicon–oxygen covalent bonds that need to be broken throughout the structure before melting occurs. At very high temperatures, silica generally occurs in different crystalline forms, such as tridymite, coesite and crystobalite, exhibiting a high refractive index and optical absorbance. Precisely, this is the same reason why it is hard in nature also. Next, if the solubility aspect is taken into account, silica exhibits low solubility in water as well as other organic solvents. The poor solubility of silica stems from the fact that there are absolutely no possible attractions between the solvent molecules and the silicon or oxygen atoms that could overcome the covalent bonds in its giant structure.

2.5.2 Chemical properties

As already specified earlier, SiO_2 has covalent bonding and forms a network-like structure. When silicon is exposed to oxygen under ambient reaction conditions, a very thin layer of silicon oxide (native oxide) is formed [71]. For obtaining well-controlled layers of silicon dioxide on silicon, higher temperature and alternate environments are utilized. It is a well-known fact that the extent of chemical modification of silica such as grafting of organo-functional groups would depend on the concentration of silanol groups per gram of silica. In the case of SiO_2 nanoparticles of size smaller than 5 nm, it has been found that more than half of the atoms are present on its surface which means that the surface would contain one or more silanol groups. The concentration of silanol group increases with the decrease in the particle size which is related to specific surface area. However, if a decrease in the concentration of silanol group with the decrease in the particle size is observed, then probably the nanoparticles would be chemically reactive, thus suitable for catalytic applications.

2.5.3 Thermal properties

SiO_2 has often been used in the formation of a thin layer in many cases of interest for the thermal management of semiconductors. However, unfortunately, the thermal conductivity (TC) of SiO_2 is about a few orders of magnitude less than Si. Hence, even the influence of a thin layer can be significant. For instance, in the case of silicon-on-insulator, thicker oxide layers would result in faster devices; however, this would be at the cost of a decrease in reliability because of higher temperature. Literature reports have revealed that there is a need to repeat the measurements because the thermal property of the thin film may vary significantly from that of the bulk. Actually, the effective TC depends upon how the layer has been created and also on its thickness. A recent report on thermally grown SiO_2 showed a clear dependence of TC on the oxide thickness [72].

2.5.4 Optical properties

Combined spectroscopic ellipsometry and UV-visible spectroscopy have been employed for studying the interband optical properties of crystalline quartz and amorphous SiO_2. It was found that the optical properties of both exhibit similar exciton and interband transitions within the range of about 1.5–4.2 eV; however, the crystalline SiO_2 shows a larger transition strength and index of refraction. Crystalline SiO_2 exhibits sharper features in the interband transition strength spectrum in comparison to the amorphous SiO_2 as the energy of absorption edge of crystalline SiO_2 is 1 eV higher. The optical properties of SiO_2 and crystalline SiO_2 were observed by Philipp experimentally in the energy range of about 0–26 eV. Later on, several studies have confirmed that optical spectra in all the SiO_2 phases with 4:2 coordination are similar. Moreover, optical properties of both crystalline and amorphous SiO_2 are important as high-purity SiO_2 crystals and glasses are important optical materials and form the basis for different optical elements [73].

2.5.5 Electrical properties

SiO_2 is an electrically insulating material as it possesses very high band gap energy, which means that the distance between the conduction and valence bands are so large that there is almost negligible possibility for the recombination of electrons and hence practically SiO_2 cannot show conduction properties [74].

2.6 Toxicology of SNPs

During the recent years, global commercialization and the large-scale industrial production of the SiO_2 NPs have resulted in the increased risk of human exposures at workplaces [75]. Also, in light of efforts to use nanomaterial in medical applications, these NPs could also be intentionally introduced into the human body for disease diagnosis and treatments. Such growing potential for exposure has raised a serious concern regarding their safety and potential adverse health effects. Studies have revealed that the SNPs generally exhibit acute toxic effects *in vitro* and *in vivo*. The data on chronic effect of these NPs are rather conflicting and not sufficient to draw any major conclusion. Also, because of the lack of realistic exposure and epidemiological data, at this moment, the translation to human health effects is impossible. The current need of the hour is to carry out detailed investigations on the SNP bioaccumulation/ bioavailability and their long-term consequences *in vivo*.

2.7 Conclusion

The last few decades have marked significant research breakthroughs in the design and synthesis of SNP-based catalytic systems which have emerged as excellent candidates for heterogeneous catalysis. In conclusion, we have described the different types of SNPs, and the various approaches involved in the synthesis of SNPs have been discussed in detail. In this chapter, particular attention has been devoted toward recent and sustainable protocols for the synthesis of SiO_2 NPs, including Stöber method (sol–gel approach), microemulsion, hydrothermal synthesis, ball milling, flame-spray pyrolysis and chemical vapor deposition. Among these, Stöber method has attracted the synthetic pursuit of researchers as it is scalable, straightforward and controllable. Additionally, we have also mentioned the detailed account of all previously published reports on the synthesis of SNPs. Furthermore, the diverse properties of SNPs have been highlighted, which include physical, chemical, thermal, optical and electrical properties. These properties basically depend upon their size, shape, composition and interfacial interaction. However, recent studies reveal that the inhalation of SNPs could cause serious toxic and adverse health effects. In contrast, their direct impact on human health and environment still require further investigation or data analysis to draw any major or final conclusion. We envision that many new synthetic methods will be developed to carry out the synthesis under safe and economical reaction conditions, such as inexpensive source of silica, shortened reaction time, economical hydrothermal treatment and scalable productions, and to avoid the use of highly alkaline or acidic wastewater. In the upcoming chapters, we will discuss the surface modification and functionalization of SNPs, and its utilization as a support for various organic transformations.

References

[1] L. L. Chng, N. Erathodiyil, J. Y. Ying, *Accounts of Chemical Research* **2013**, *46*, 1825–1837.

[2] D. Zhang, X. Du, L. Shi, R. Gao, *Dalton Transactions* **2012**, *41*, 14455–14475.

[3] Y. G. Guo, J. S. Hu, L. J. Wan, *Advanced Materials* **2008**, *20*, 2878–2887.

[4] G. Yang, C. Zhu, D. Du, J. Zhu, Y. Lin, *Nanoscale* **2015**, *7*, 14217–14231.

[5] J. Gu, Y. W. Zhang, F. F. Tao, *Chemical Society Reviews* **2012**, *41*, 8050–8065.

[6]   S. B. Kalidindi, B. R. Jagirdar, *ChemSusChem* **2012**, *5*, 65–75.
[7]   H. H. Kung, M. C. Kung, *Nanotechnology in Catalysis*, Springer, NY, **2007**, pp. 1–11.
[8]   R. Jin, *Nanotechnology Reviews* **2012**, *1*, 31–56.
[9]   T. Chen, V. O. Rodionov, *ACS Catalysis* **2016**, *6*, 4025–4033.
[10] M. C. Daniel, D. Astruc, *Chemical Reviews* **2004**, *104*, 293–346.
[11] H. Lee, *RSC Advances* **2014**, *4*, 41017–41027.
[12] S. Schauermann, N. Nilius, S. Shaikhutdinov, H. J. Freund, *Accounts of Chemical Research* **2012**, *46*, 1673–1681.
[13] M. B. Gawande, R. Zboril, V. Malgras, Y. Yamauchi, *Journal of Materials Chemistry A* **2015**, *3*, 8241–8245.
[14] J. Wang, H. Gu, *Molecules* **2015**, *20*, 17070–17092.
[15] C. W. Lim, I. S. Lee, *Nano Today* **2010**, *5*, 412–434.
[16] R. K. Sharma, S. Sharma, S. Dutta, R. Zboril, M. B. Gawande, *Green Chemistry* **2015**, *17*, 3207–3230.
[17] R. Nandanwar, P. Singh, F. Zia Haque, *Material Science Research* **2013**, *10*, 85–93.
[18] T. Suteewong, H. Sai, J. Lee, M. Bradbury, T. Hyeon, S. M. Gruner, U. Wiesner, *Journal of Materials Chemistry* **2010**, *20*, 7807–7814.
[19] V. Polshettiwar, D. Cha, X. Zhang, J. M. Basset, *Angewandte Chemie International Edition* **2010**, *49*, 9652–9656.
[20] N. Linares, E. Serrano, M. Rico, A. M. Balu, E. Losada, R. Luque, J. Garcia-Martinez, *Chemical Communications* **2011**, *47*, 9024–9035.
[21] R. Luque, A. M. Balu, J. M. Campelo, M. D. Gracia, E. Losada, A. Pineda, A. A. Romero, J. C. Serrano-Ruiz, *Catalysis* **2014**, *24*, 253–280.
[22] R. K. Sharma, S. Sharma, G. Gaba, S. Dutta, *The Journal of Materials Science* **2016**, *51*, 2121–2133.
[23] E. F. Vansant, P. Van Der Voort, K. C. Vrancken, *Characterization and Chemical Modification of the Silica Surface*, Vol. 93, 1st edn., Studies in surface science and catalysis, Elsevier, **1995**, pp. 59–77.
[24] R. P. Bagwe, L. R. Hilliard, W. Tan, *Langmuir* **2006**, *22*, 4357–4362.
[25] Y. Li, B. C. Benicewicz, *Macromolecules* **2008**, *41*, 7986–7992.
[26] M. Chidambaram, R. Manavalan, K. Kathiresan, *The Journal of Pharmacy and Pharmaceutical Sciences* **2011**, *14*, 67–77.
[27] http://nanosikkerhed.nu/wp-content/uploads/2014/12/Rambabu-SilicaCategorization.pdf.
[28] D. Kolbe, *Komplexchemische Verhalten der Kieselsaure*. Dissertation, Friedrich-Schiller University Jena, **1956**.
[29] W. Stöber, A. Fink, E. Bohn, *The Journal of Colloid and Interface Science* **1968**, *26*, 62–69.
[30] R. Nandanwar, P. Singh, F. Z. Haque, *Material Science Research India* **2013**, *10*, 85–92.
[31] I. A. Rahman, V. Padavettan, *Journal of Nanomaterials* **2012**, *8*, 1–15.
[32] A. K. Van Helden, J. W. Jansen, A. Vrij, *The Journal of Colloid and Interface Science* **1981**, *81*, 354–368.
[33] A. P. Philipse, *Colloid and Polymer Science* **1988**, *266*, 1174–1180.

[34] A. P. Philipse, A. Vrij, *The Journal of Colloid and Interface Science* **1989**, *128*, 121–136.

[35] G. H. Bogush, C. F. Zukoski IV, *The Journal of Colloid and Interface Science* **1991**, *142*, 1–18.

[36] G. H. Bogush, C. F. Zukoski IV, *The Journal of Colloid and Interface Science* **1991**, *142*, 19–34.

[37] A. Van Blaaderen, J. Van Geest, A. Vrij, *The Journal of Colloid and Interface Science* **1992**, *154*, 481–501.

[38] A. Van Blaaderen, A. Vrij, *The Journal of Colloid and Interface Science* **1993**, *156*, 1–18.

[39] T. Yokoi, J. Wakabayashi, Y. Otsuka, W. Fan, M. Iwama, R. Watanabe, T. Okubo, *Chemistry of Materials* **2009**, *21*, 3719–3729.

[40] O. M. Sadek, S. M. Reda, R. K. Al-Bilali, *Advances in Nanoparticles* **2013**, *2*, 165–175.

[41] D. D. Lovingood, J. R. Owens, M. Seeber, K. G. Kornev, I. Luzinov, *ACS Applied Materials Interfaces* **2012**, *4*, 6875–6883.

[42] D. Tang, W. Zhang, Y. Zhang, Z. A. Qiao, Y. Liu, Q. Huo, *The Journal of Colloid and Interface Science* **2011**, *356*, 262–266.

[43] J. C. Song, F. F. Xue, Z. Y. Lu, Z. Y. Sun, *Chemical Communications* **2015**, *51*, 10517–10520.

[44] J. H. Schulman, W. Stoeckenius, L. M. Prince, Mechanism of formation and structure of micro emulsions by electron microscopy, *The Journal of Physical Chemistry* **1959**, *63*, 1677–1680.

[45] M. A. Malik, M. Y. Wani, M. A. Hashim, *The Arabian Journal of Chemistry* **2012**, *5*, 397–417.

[46] F. J. Arriagada, K. Osseo-Asare, *The Journal of Colloid and Interface Science* **1999**, *211*, 210–220.

[47] K. S. Finnie, J. R. Bartlett, C. J. A. Barbé, L. Kong, *Langmuir* **2007**, *23*, 3017–3024.

[48] C. H. Lin, J. H. Chang, Y. Q. Yeh, S. H. Wu, Y. H. Liu, C. Y. Mou, *Nanoscale* **2015**, *7*, 9614–9626.

[49] Y. Han, J. Jiang, S. S. Lee, J. Y. Ying, *Langmuir* **2008**, *24*, 5842–5848.

[50] J. Pak, H. Yoo, *Journal of Materials Chemistry A* **2013**, *1*, 5408–5413.

[51] K. Byrappa, M. Yoshimura, *Handbook of Hydrothermal Technology: A Technology for Crystal Growth and Materials Processing*, Norwich, **2001**.

[52] A. Rabenau, *Angewandte Chemie International Edition* **1985**, *24*, 1026–1040.

[53] L. Gu, A. Zhang, K. Hou, C. Dai, S. Zhang, M. Liu, X. Guo, *Microporous and Mesoporous Materials* **2012**, *152*, 9–15.

[54] Q. Yu, P. Wang, S. Hu, J. Hui, J. Zhuang, X. Wang, *Langmuir* **2011**, *27*, 7185–7191.

[55] N. Qisti, N. S. Indrasti, *Materials Science and Engineering C* **2016**, *162*, 012036.

[56] S. Iravani, *Green Chemistry* **2011**, *13*, 2638–2650.

[57] V. V. Makarov, A. J. Love, O. V. Sinitsyna, S. S. Makarova, I. V. Yaminsky, M. E., Taliansky, N. O. Kalinina, *Acta Naturae* **2014**, *6*.

[58] N. O. San, C. Kurşungöz, Y. Tümtaş, Ö. Yaşa, B. Ortac, T. Tekinay, *Particuology* **2014**, *17*, 29–35.

[59] S. Rovani, J. J. Santos, P. Corio, D. A. Fungaro, *ACS Omega* **2018**, *3*, 2618–2627.

[60] https://www.pcimag.com/articles/100364-biogenic-silica-harvested-from-r ice-hulls.

[61] U. Vijayalakshmi, V. Vaibhav, M. Chellappa, U. Anjaneyulu, *Journal of the Indian Chemical Society* **2015**, *92*, 675–678.

[62] C. L. Nistor, R. Ianchis, M. Ghiurea, C. A. Nicolae, C. I. Spataru, D. C. Culita, J. Pandele Cusu, V. Fruth, F. Oancea, D. Donescu, *Nanomaterials* **2016**, *6*, 9.

[63] A. de la Hoz, A. Díaz-Ortiz, P. Prieto, *Alternative Energy Sources for Green Chemistry*, Royal Society of Chemistry **2016**, pp. 1–33.

[64] A. Santiago, A. González, J. J. Iruin, *The Journal of Sol-Gel Science and Technology* **2012**, *61*, 8.

[65] E. Mily, A. González, J. J. Iruin, L. Irusta, M. J. Fernández-Berridi, *The Journal of Sol-Gel Science and Technology* **2010**, *53*, 667–672.

[66] P. Ubat, *American Journal of Analytical Chemistry* **2016**, *20*, 1382–1389.

[67] D. Webb, T. F. Jamison, *Chemical Science* **2010**, *1*, 675–680.

[68] L. Malet-Sanz, F. Susanne, *Journal of Medicinal Chemistry* **2012**, *55*, 4062–4098.

[69] A. B. Corradi, F. Bondioli, A. M. Ferrari, B. Focher, C. Leonelli, *Powder Technology* **2006**, *167*, 45–48.

[70] N. Gharehbash, A. Shakeri, *Oriental Journal of Chemistry* **2015**, *31*, 207–212.

[71] M. Morita, T. Ohmi, E. Hasegawa, M. Kawakami, M. Ohwada, *Journal of Applied Physics* **1990**, *68*, 1272–1281.

[72] C. Huang, Z. Lin, Y. Feng, X. Zhang, G. Wang, *The European Physical Journal* **2015**, *130*, 239.

[73] I. A. Rahman, P. Vejayakumaran, C. S. Sipaut, J. Ismail, C. K. Chee, *Materials Chemistry and Physics* **2009**, *114*, 328–332.

[74] M. Barisik, S. Atalay, A. Beskok, S. Qian, *The Journal of Physical Chemistry C* **2014**, *118*, 1836–1842.

[75] W. Lin, Y. W. Huang, X. D. Zhou, Y. Ma, *Toxicology and Applied Pharmacology* **2006**, *217*, 252–259.

Chapter 3

Silica-Encapsulated Magnetic Nanoparticles

Yukti Monga, Radhika Gupta, Rashmi Gaur
and Rakesh Kumar Sharma*
*rksharmagreenchem@hotmail.com

3.1 Introduction

3.1.1 Magnetic nanoparticles

Magnetic nanoparticles (MNPs) have fascinated numerous researchers working in various fields of science and technology [1]. The unique properties of these nanoparticles (NPs), resulting from their high surface area-to-volume ratio, make them highly suitable for different useful applications, including data storage [2], as ferrofluids [3], magnetic refrigeration systems [4], biotechnology (protein purification, medical imaging and drug delivery) [5], catalysis [6] and environmental remediation [7]. Some features of these particles are quite revolutionary, such as controllable size and shape, high thermal stability and easy manipulation, which results in high surface reactivity, better kinetics and finest efficiency (Figure 3.1). Besides, facile separation, with the aid of an external magnet, provides green and sustainable feature to their potentials.

---

*Corresponding author.

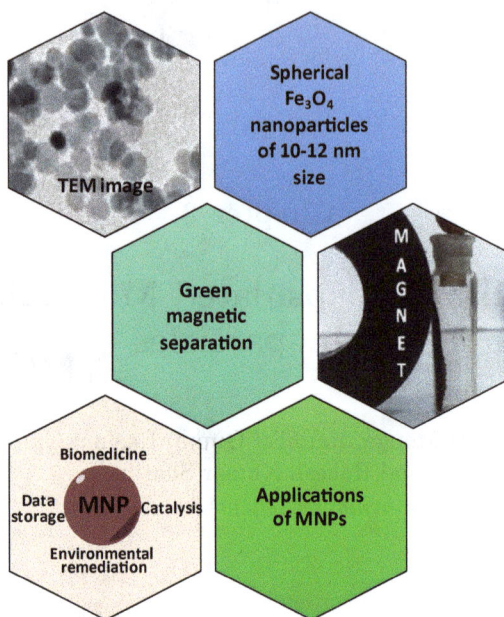

Figure 3.1. Properties and applications of MNPs.

3.1.2 Magnetite as the ultimate catalytic support

Recyclability of catalysts in organic reactions is the bottleneck for any industrial application. One of the keynotes for this problem is the green recovery of the catalytic species. "Magnetically driven separation" has opened new avenues in this regard as it offers the opportunity to make manufacturing processes simpler, safer and less time-consuming [6]. It also avoids the consequences of filtration steps, such as catalyst mass loss, use of additional solvents and consequent generation of organic residues. Subsequently, MNPs have proved themselves as excellent catalytic supports and thereby contribute in the development of green and pollution abatement catalysts. They are highly dispersible in reaction media, which enhances the catalytic rate and efficiency. They also permit the feasibility of altering the sites and their morphology, so that they can be designed for a specific application in catalysis. Hence, they meet most of the requirements of being an ideal and capable catalytic support.

Magnetite, an inverse spinel
Fe_3O_4, or $Fe(FeO_2)_2$

● octahedral Fe(II, III)

● tetrahedral Fe(III)

● oxide anion

Magnetic Crystal

Crystal system: Isometric hexoctahedral

Crystal Habit
A Site tetrahedral Fe (III)
B Site Octahedral (II, III)

Figure 3.2.   Crystal structure of magnetite (adapted from Ref. [9]).

Various types of MNPs are known, such as pure metal-based (Fe, Co), alloys (CoPt$_3$, FePt) and spinel type ferromagnets (AB$_2$O$_4$; A=Mg, Mn, Co, Fe; B=Fe) [1c]. Among all, magnetite nanoparticles (Fe$_3$O$_4$ NPs) are most widely used because of their high saturation magnetization value [8]. Fe$_3$O$_4$ nanoparticles are black, metallic mineral particles containing both Fe(II) and Fe(III) ions. The crystal structure of magnetite is inverse spinel with a unit cell containing 32 oxygen atoms in a face-centered cubic structure and unit cell edge length of 0.839 nm (Figure 3.2) [9]. In this structure, Fe(II) ions and Fe(III) ions occupy the octahedral sites and the other half Fe(III) ions occupy the tetrahedral sites. The crystal forms of magnetite include octahedron and rhombodecahedron.

Herein, we describe various methods for the synthesis of Fe$_3$O$_4$ NPs. The main focus of this chapter is to discuss the need, benefits and methods of silica encapsulation of MNPs. Different regulations for the formation of uniform silica coat on MNPs have also been described. Special attention is given to the properties of silica-coated magnetic nanoparticles (SMNPs) which promote their use as an ideal catalytic support in various organic transformations.

3.2   Methods for the Synthesis of Fe$_3$O$_4$

Since the last few decades, considerable amount of research has been done on the synthesis of MNPs. Numerous publications have described synthetic routes to control the size, uniformity, crystallinity

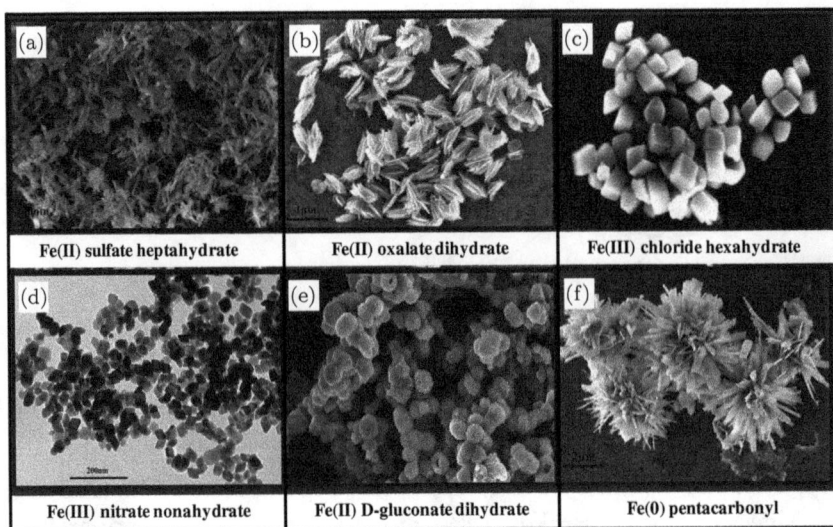

Figure 3.3. SEM images of iron oxide nanoparticles having different morphologies: (a) nanorod, (b) nanohusk, (c) distorted cubes, (d) nanocubes, (e) porous spheres and (f) self-oriented flowers (adapted from Ref. [10]).

and stability of NPs. Recently, Sayed and Polshettiwar reported that besides size, shape also plays an integral role in determining the properties of NPs [10]. They synthesized six different morphologies of iron oxide nanoparticles using the same synthetic protocol and by simply changing the iron precursor. Figure 3.3 displays the SEM images of the obtained NPs (using the precursors written below the respective image).

Table 3.1 summarizes various popular approaches that have been utilized for the synthesis of MNPs.

3.3 Coating of MNPs

MNPs have emerged as new support material for catalyst immobilization. But naked MNPs alone are not very successful in industrial applications. They possess strong magnetic interactions which cause aggregation of particles [1c, 1e]. When exposed to air and other strong chemical environments, their original structure may get changed resulting in the alteration of magnetic properties. They can undergo

Table 3.1. Approaches for the synthesis of Fe_3O_4 nanoparticles.

Process	Approach	Techniques used	Reagents and conditions	Ref.
Physical methods	Gas-phase deposition	Laser vaporization, thermal vaporization, arc discharge, plasma vaporization and solar energy-induced evaporation	Iron precursors; vary as per the technique	[1a]
	Electron beam lithography	Electron beam lithograph	—	[11]
	Ball milling method (a bulk method)	Ball mill	Ambient conditions	[12]
Chemical methods	Co-precipitation	Reduction	Metal salts, base; 20–90°C, pH < 9	[13]
	Microemulsion	Formation of micelles or inverse micelles	Soluble metal salt, surfactant, base; 20–50°C, complicated conditions	[14]
	Thermal decomposition	Decomposition of organometallic precursors: $Fe(cup)_3$ (cup = N-nitrosophenylhydroxylamine), $Fe(acac)_3$, $Fe(CO)_5$	Organometallic precursor, organic solvents, surfactants (oleylamine, hexadecylamine); inert atmosphere, 100–320°C	[15]
	Hydrothermal method	Hydrolysis at high temperature	Ferric acetylacetonate as the sole iron source, poly(acrylic acid) as stabilizer; high temperature (130–250°C) and pressure (0.3–4 MPa)	[16]

(Continued)

Table 3.1. (*Continued*)

Process	Approach	Techniques used	Reagents and conditions	Ref.
	Polyol	—	Polyols (ethylene glycol, diethylene glycol), base, solvents, metal salts; very high temperature (280°C), unfavorable conditions	[17]
	Solvothermal method	—	Ethylene glycol, oleic acid, hexadecylamine; 30–40°C; DCM as solvent	[18]
	Electrochemical	Electro-oxidation	Amine surfactants, supporting electrolyte, coating agent; use of electrochemical cell, distance between anode and cathode is kept as per the requirement	[19]
	Continuous flow technique	Continuous tubular flow reactor	Iron precursors, organic solvents; high temperature and pressure conditions	[20]
Biological methods	Using biological methods	Using microorganisms	Vary according to the microorganism	[21]
Green methods	Using green methods	Using renewable resources	Ambient conditions (generally room temperature)	[22]

rapid biodegradation when directly exposed to the biological system. They can also cause cytotoxicity due to intracellular dissolution and can strongly damage the DNA. It has been found that ferric ions can be released from these nanoparticles which are actually responsible for the toxicity mechanism. These ions can react with H_2O_2 produced by the mitochondria resulting in the formation of highly reactive hydroxyl radicals through the Fenton reagent [23]. Their applicability could be increased by coating their surface to protect the magnetic core and by increasing the surface reactivity. It has been demonstrated that the formation of a passive coat of inert material on the surface of iron oxide nanoparticles could help prevent their aggregation in liquid and improve their chemical stability [24]. Table 3.2 demonstrates different materials by which MNPs can be coated.

3.4 Coating of MNPs with Silica

Out of various protective coatings, silica delivers several key advantages (Figure 3.4) [31]. It has been found that MNPs have strong surface affinity toward silica. MNPs can be directly coated with silica without the need of any primer, which is generally required to promote the decomposition and adhesion of silica. This silica coating prevents magnetic talking, possible decomposition due to surrounding environment and oxidation of core material. The main benefit of silica coating is the improved ability to form strong covalent bonds with various functionalization molecules. Another advantage is that its surface often terminated with silanol groups that can further react with various coupling agents to allow the covalent immobilization of specific ligands on the surface of the MNPs and thereby reduce their desorption [32]. Beyond this, silica offers some pronounced advantages over other supports such as the following: ease of attachment of functionalities, provides great resistance to organic solvents, improves the dispersity index of the particles and provides high chemical and thermal resistance. Silica coating also improves solubility by converting hydrophobic NPs into hydrophilic water-soluble particles [31, 33].

Table 3.2.   Different coating strategies on MNPs.

Coating strategies	Coating precursors	Advantages	Applications	Ref.
Polysaccharide	Starch, chitosan, dextran, alginate, pullulan, etc.	• Better biocompatibility • High chemical reactivity • High binding selectivity with target site	Imaging and gene therapy	[25]
Inert metal	Gold, silver	• Provides stability against corrosive biological conditions due to their noble nature • Gold coating provides conductivity for sensing and optical activities • Favor bioaffinity through functionalization by thiol terminal groups • Silver coating imparts antimicrobial activity	Bioseparation, DNA sensing, drug delivery, antimicrobial activity	[26]
Silica	TEOS, sodium silicate	• High chemical and thermal stability • Large surface area due to porous structure • Improved ability for functionalization	Catalysis, solid-phase adsorption	[6a, 7, 27]

Polymer and dendrimer	Polyethylene glycol, poly(ethylene imine), poly (lactide-co-glycolide), etc.	• Prevents aggregation by steric shielding • Provides colloidal stability • Provides higher density of functional end groups	Catalysis, scavenging, drug delivery	[28]
Carbon	Glucose, graphene, resorcinol–formaldehyde, etc.	• Carbonaceous shells provide higher stability in acidic/basic medium • Has porous structure and hence larger surface area	Solid-phase extraction sorbent, water treatment, energy storage, catalysis	[29]
Metal oxide	TiO_2, Al_2O_3, MgO	• Prevents magnetic core from sintering at high temperature, and thus, preserves magnetic properties	Catalysis, water remediation	[30]

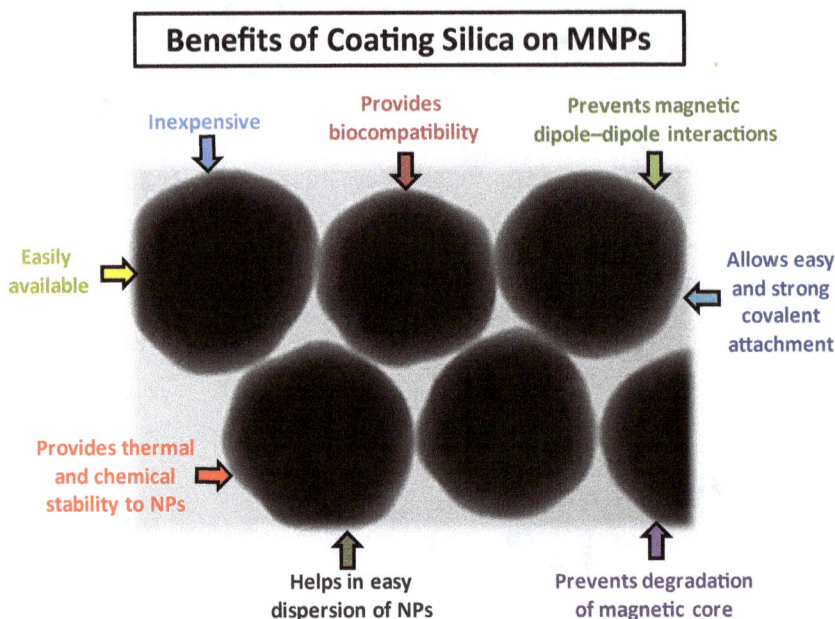

Figure 3.4.   Benefits of silica coating (adapted from Ref. [31]).

3.5   Methods of Silica Coating

There are five basic methods of silica coating on MNPs, as explained in Sections 3.5.1–3.5.5.

3.5.1   Stöber sol–gel process

The sol–gel process is a term in which dispersion of solid nanoparticles occurs in a liquid "sol" and these nanoparticles agglomerate together in order to form a continuous three-dimensional network in the liquid-phase "gel". In this method, silica phase is formed on the surface of MNPs in a basic alcohol/water mixture by continuous hydrolysis and condensation of TEOS (Scheme 3.1) [33b, 34]. Sharma *et al.* synthesized SMNPs by first activating the MNPs in HCl and then diluting them with water, ethanol and ammonia. Further the mixture was irradiated with ultrasonic vibrations [27o]. Finally, TEOS was added slowly to the above dispersion and stirred at

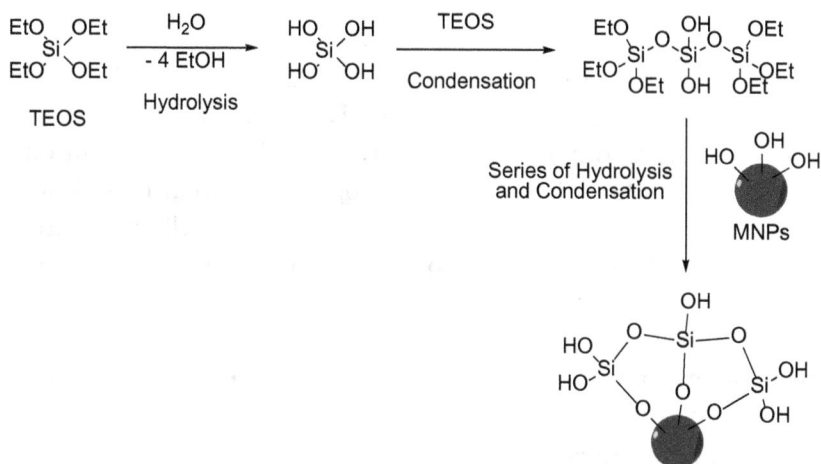

Scheme 3.1. Mechanism of silica coating on the surface of MNPs.

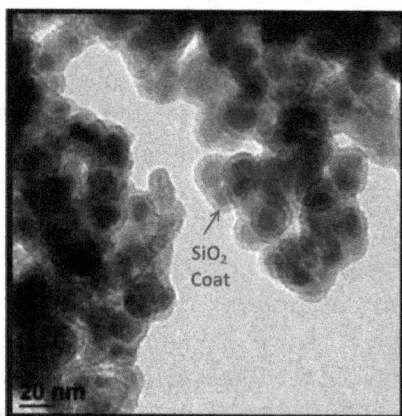

Figure 3.5. TEM image of SMNPs prepared by the sol–gel process (2 g MNPs, 160 mL EtOH, 40 mL distilled water, 5 mL aqueous ammonia, 1 g TEOS) (adapted from Ref. [27o]).

60°C for 6 h to obtain the silica coat around MNPs. The transmission electron microscopy (TEM) image of the obtained SMNPs shows a core–shell structure with a mean diameter of 16–22 nm (Figure 3.5).

3.5.2 Silicic acid method

Liu *et al.* utilized this method for coating silica on MNPs using sodium silicate as the silica source [35]. For this, Fe_3O_4 nanoparticles were mixed with sodium silicate solution followed by the addition of HCl to adjust the pH to 6.0. The dropwise addition of HCl leads to the formation of silicic acid which condenses with the hydroxyl groups of Fe_3O_4 nanoparticles to form strong covalent bonds. Afterward, polymerization takes place *via* condensation reaction, thereby transforming silanol groups (Si–OH) to siloxane bonds (Si–O–Si). In this manner, compact silica coat grows on the surface of MNPs.

3.5.3 Aerosol pyrolysis

It is a method where SMNPs are prepared in an aerosol reactor by nebulizing solutions of precursor salts (silicon alkoxides and metal compound) and further pyrolyzing in a flame environment [33b].

There are two types of aerosol pyrolysis: spray and laser.

Spray pyrolysis: In spray pyrolysis, solid particles are obtained by spraying a precursor solution into a series of reactors where the aerosol droplets undergo evaporation of the solvent, solute condenses within the droplet, followed by drying and finally pyrolysis of the precipitated particles takes place at high temperature. The final diameter of the resulting spherical particles can be predetermined from that of the original droplets. A schematic representation of the aerosol spray reactor is shown in Figure 3.6 [32, 36].

Tartaj *et al.* reported the direct synthesis of silica-coated γ-Fe_2O_3 hollow spherical particles by the spray pyrolysis of methanol solutions containing iron ammonium citrate and TEOS (Figure 3.7) [36b]. A possible formation mechanism involves rapid evaporation of methanol which favors surface precipitation of components, i.e. formation of hollow spheres (as iron ammonium citrate is less soluble in methanol than TEOS, the iron salt precipitates in a faster manner, thereby promoting the formation of solid iron shell). Further, the enrichment of TEOS occurs at the surface of the solid iron shell. Finally, the thermal decomposition of precursors produces the silica-coated γ-Fe_2O_3 hollow spheres.

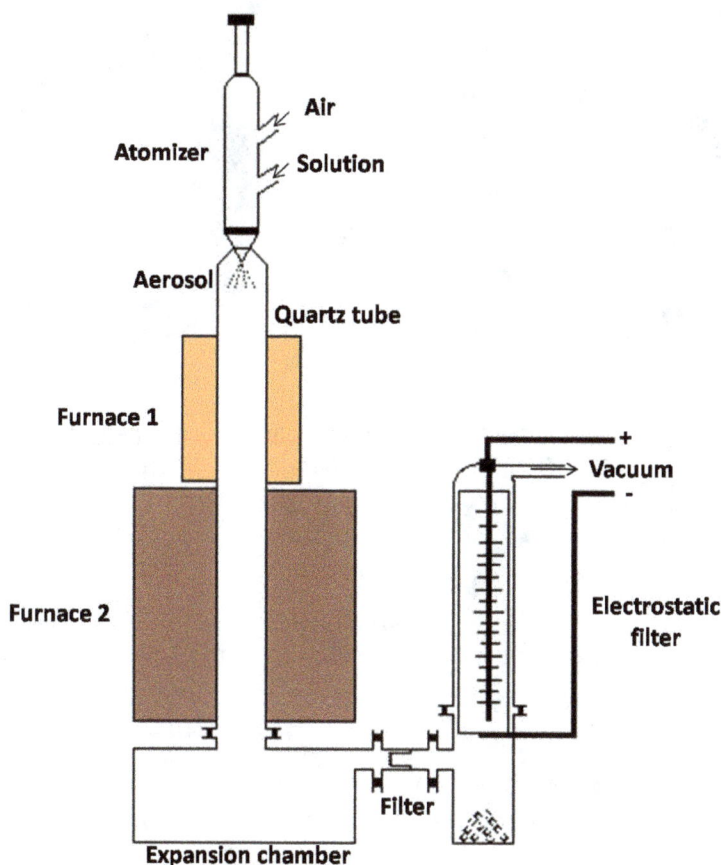

Figure 3.6. Schematic representation of the aerosol spray reactor.

Laser pyrolysis: This method involves heating of a dilute mixture of vaporized precursors with a continuous-wave carbon dioxide laser, which initiates and sustains particle nucleation. Final size of the particles depends on the aggregation of the initial nuclei. A representation of laser pyrolysis flowchart is given in Figure 3.8 [37].

Using this technique, Bomatí-Miguel *et al.* prepared SMNPs [38]. They nebulized a mixture of ferrocene and TEOS using ultrasound and the cloud so formed was heated using horizontal CO_2 laser beam. When the experiment was carried out under inert or

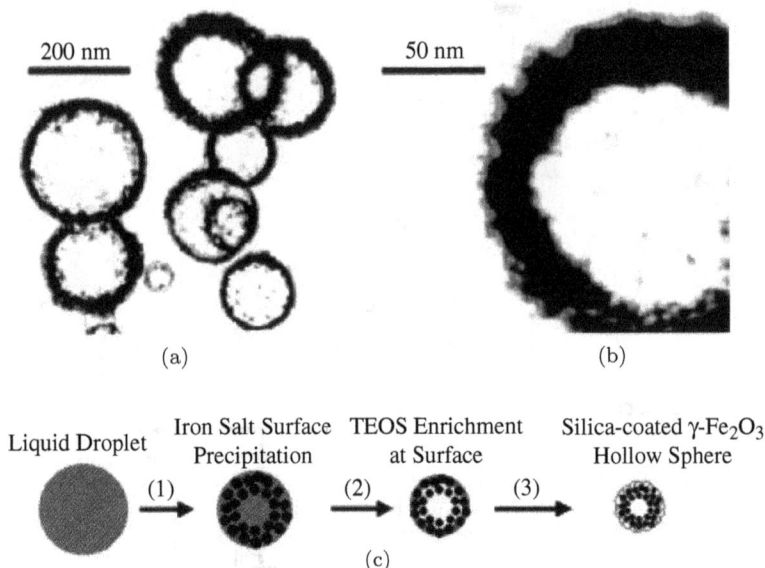

Figure 3.7. (a), (b) TEM images and (c) formation mechanism of silica-coated γ-Fe$_2$O$_3$ hollow spherical particles formed by spray pyrolysis method (adapted from Ref. [36b]).

reduced conditions, SiO$_2$ composite containing α-Fe was obtained (Figure 3.9(a)) but upon the introduction of air, a mixture of SiO$_2$-encapsulated α-Fe and spinel iron oxide was formed (Figure 3.9(b)).

3.5.4 Arc discharge method

It has been reported that silica coating can also be done by the means of modified arc discharge method [39]. The arc discharge furnace consists of a cylindrical chamber and two electrodes: a stationary hollow iron anode containing finely powdered silica precursor (zeolite/metallic silicon) and a movable tungsten cathode (Figure 3.10). The entire system is water-cooled and the chamber is filled with He gas after evacuation. An arc is produced between the electrodes which ionizes the He plasma. When the temperature of the system reaches 3000°C, the iron and silica powder sublime and then condense in the colder area of the chamber (iron nuclei sublime

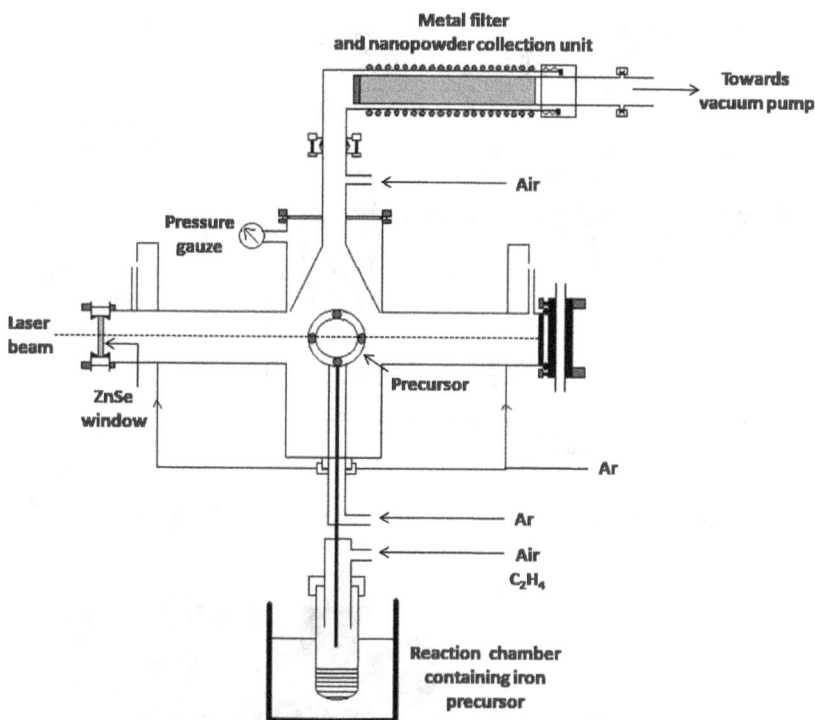

Figure 3.8.   Schematic representation of the laser pyrolysis.

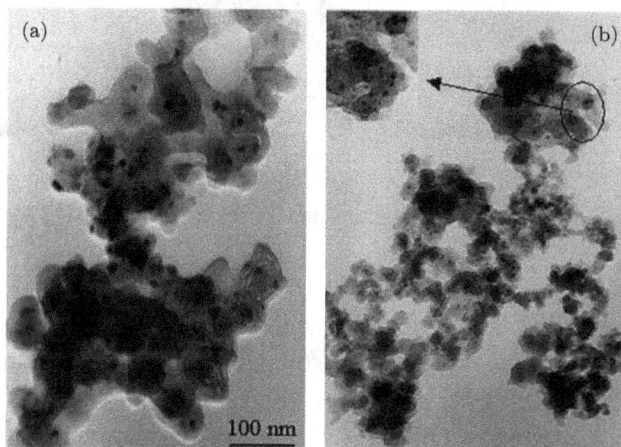

Figure 3.9.   TEM image of SiO_2 composite containing (a) only iron and (b) some iron oxide (adapted from Ref. [38]).

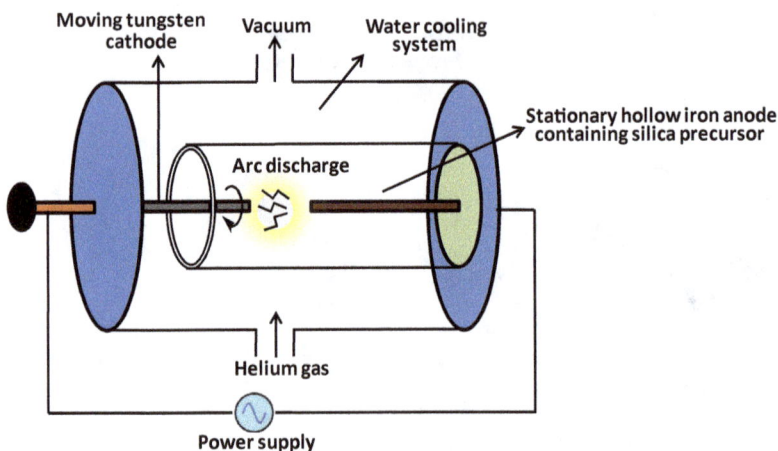

Figure 3.10.   Schematic representation of arc discharge furnace.

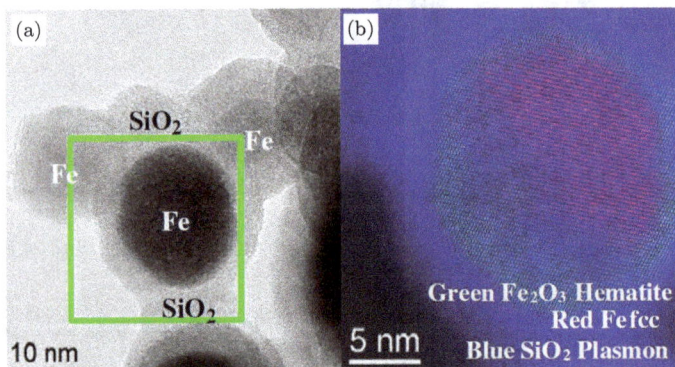

Figure 3.11.   (a) High-resolution TEM (HRTEM) and (b) energy-filtering TEM (EFTEM) images of silica-coated iron nanoparticles formed using arc discharge method (adapted from Ref. [39]).

faster than silica thereby forming silica shell over iron core). Finally, the nanoparticles were collected from the walls of the chamber. The HRTEM and EFTEM images of the so-formed silica-coated iron nanoparticles are shown in Figure 3.11. The saturation magnetization value of the formed core/shell particles is sufficiently high (over 160 emu/g).

3.5.5   Microemulsion method

Reverse microemulsion is a kind of "water-in-oil (w/o)" microemulsion where water is dispersed in the hydrocarbon phase, and upon the addition of a surfactant, reverse micelles are obtained [40]. Using this method, Lee and co-workers coated silica on FePt nanocrystals [41]. For this, oleic acid-/oleylamine-capped FePt nanocrystals were dispersed in cyclohexane and then injected into the cycloheaxne/Igepal solution. Further, NH_4OH aqueous solution was added followed by the addition of TEOS. The mixture was stirred for 72 h to obtain FePt@SiO$_2$ (Figure 3.12).

Vogt *et al.* suggested a ligand exchange mechanism for the formation of silica coating over hydrophobic MNPs [42]. First, ligand exchange occurs between the surface hydrophobic ligands of MNPs and partially hydrolyzed TEOS species. In this manner, MNPs get transferred from the oil phase to the water phase, which is essential for the formation of core–shell structure. Further, silica shell grows on the surface of water phase-transferred MNPs due to the continuous interaction of hydrolyzed TEOS of the oil phase with

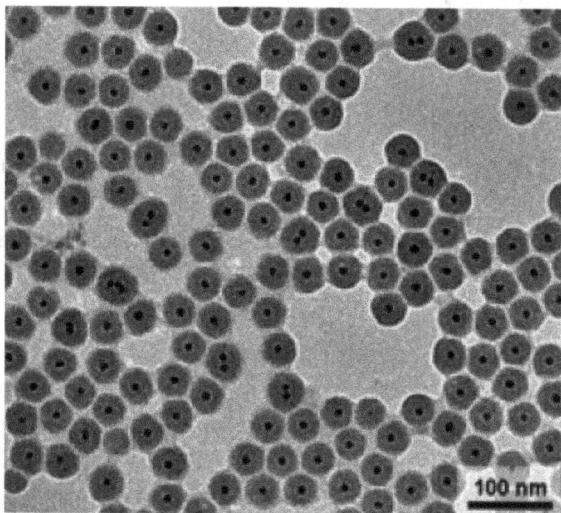

Figure 3.12.   TEM image of FePt@SiO$_2$ (adapted from Ref. [41]).

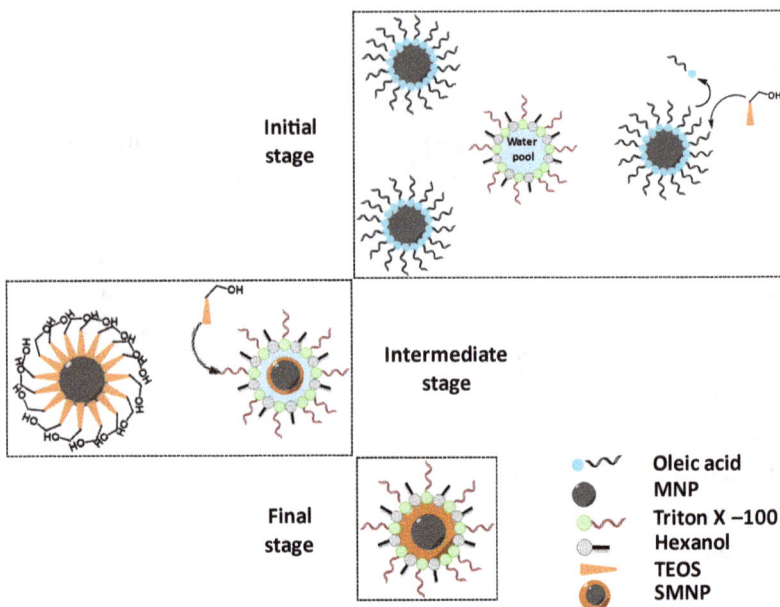

Figure 3.13. Formation mechanism of SMNPs using reverse microemulsion method.

the inverse micelles. This results in the formation of silica shell-coated MNPs (Figure 3.13).

3.6 Silica-Coating Regulations

For practical applications, it is required that every nanoparticle should be coated with a continuous, homogeneous and thin silica shell without the formation of free-core silica or multicore silica, thereby not impairing the magnetic properties. Ding *et al.* reported a systematic study on the silica-coating regulation of iron oxide nanoparticles *via* the reverse microemulsion method [43]. They explained that the uniform coating of silica on Fe_3O_4 nanoparticles with a single core and without core-free or multiple-core silica *via* reverse microemulsion method is dependent on the molar ratio of ammonia to Igepal CO-520 (denoted as R), the size and loading content of hydrophobic Fe_3O_4 nanoparticles, amount of ammonia and content of TEOS. Table 3.3 and Figure 3.14 show that the

Table 3.3.   Effect of content and size of Fe_3O_4 NPs for the formation of core-free or multicore-free $Fe_3O_4@SiO_2$ NPs.

Sample name	Content of Fe_3O_4 (mg)	Size of Fe_3O_4 NPs (nm)	Formation of core-free silica
A1	0.8	8.8	No
A2	0.8	12.2	Yes
A3	1.6	12.2	Yes
A4	2.0	12.2	No

Note: Ratio of Igepal CO-520 (g) to ammonia (μL) is 1.36:100, TEOS 75 μL.

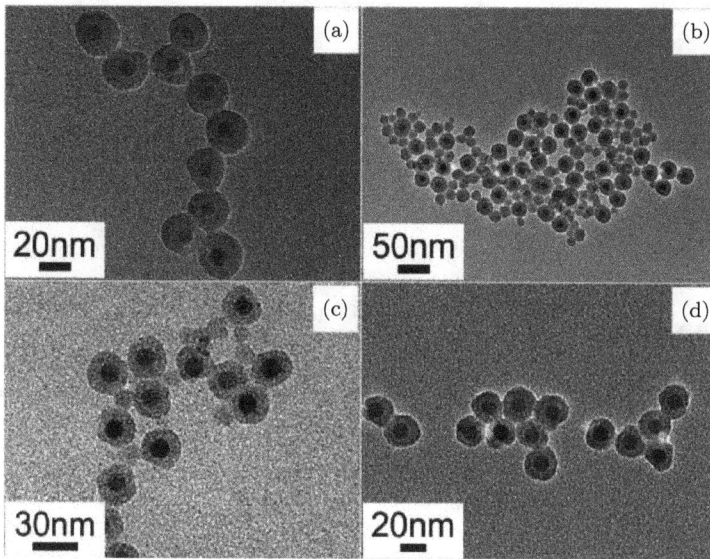

Figure 3.14.   TEM images of $Fe_3O_4@SiO_2$ NPs of samples. (a) A1, (b) A2, (c) A3 and (d) A4 (adapted from Ref. [43]).

coating regulation applied on 8.8 nm Fe_3O_4 nanoparticles cannot be applied on 12.2 nm Fe_3O_4 nanoparticles. To obtain core-free Fe_3O_4 (12.2 nm)$@SiO_2$ NPs, the loading content of Fe_3O_4 must be increased from 0.8 to 2.0 mg. It happened because in the reverse microemulsion method, the number and size of aqueous domains are dependent on R. With a fixed R value, the size and number of aqueous domains will be constant. If the number of Fe_3O_4 NPs equals the number of

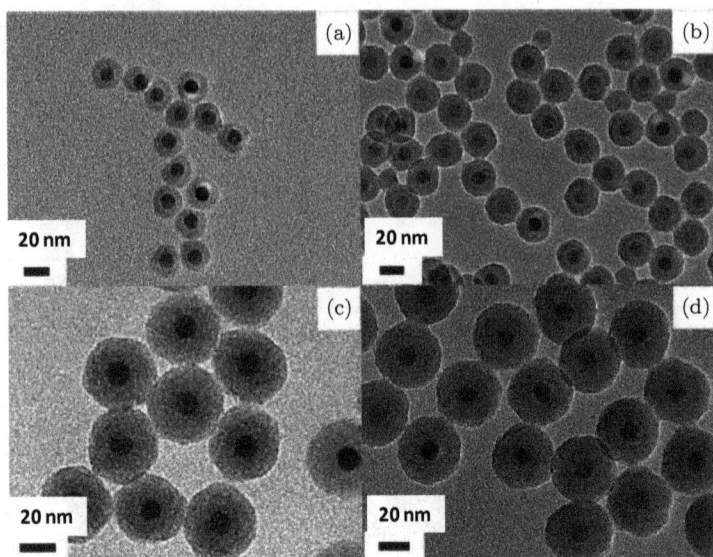

Figure 3.15. TEM images of $Fe_3O_4@SiO_2$ NPs of samples. (a) B1, (b) B2, (c) B3 and (d) B4 (adapted from Ref. [43]).

Table 3.4. Effect of amount of TEOS.

Sample name	Amount of TEOS (μL)	SiO_2 shell thickness (nm)	Formation of core-free silica
B1	75	7.0	No
B2	150	8.7	Yes
B3	300 (in 4 fractions)	13.3	No
B4	600 (in 6 fractions)	14.9	No

Note: Content of 12.2 nm Fe_3O_4 1.3 mg, ratio of Igepal CO-520 (g) to ammonia (μL) is 0.5:100.

aqueous domains, core-free or multicore-free $Fe_3O_4@SiO_2$ NPs will be formed.

It was also examined how to effectively increase the silica-coat thickness of Fe_3O_4 NPs without forming core-free silica. It was reported that the thickness of silica shell increases as the amount of TEOS increases (Figure 3.15). But this also leads to the formation of core-free silica. To avoid this, they adopted equivalently fractionated drop method (Table 3.4).

It was further explained that the content of ammonia also affects the silica thickness. Table 3.5 shows that on increasing the ammonia content from 50 to 200 μL, the silica thickness increased from 5 to 8.3 nm (Figure 3.16). This is because on increasing the content of ammonia, the size of aqueous domain will increase which would have freer aqueous space thereby resulting in the formation of more monomers and thus increasing the silica shell thickness. But this also led to the formation of core-free silica particles. To overcome

Table 3.5. Effect of amount of ammonia.

Sample name	Amount of ammonia (μL)	Amount of TEOS (μL)	SiO$_2$ shell thickness (nm)	Formation of core-free silica
C1	50	75	5.0	No
C2	100	75	7.0	No
C3	200	75	8.3	Yes
C4	200	35	6.3	No

Note: Fe$_3$O$_4$ NPs (12.2 nm) = 1.3 mg, Igepal CO-520 = 0.5 g.

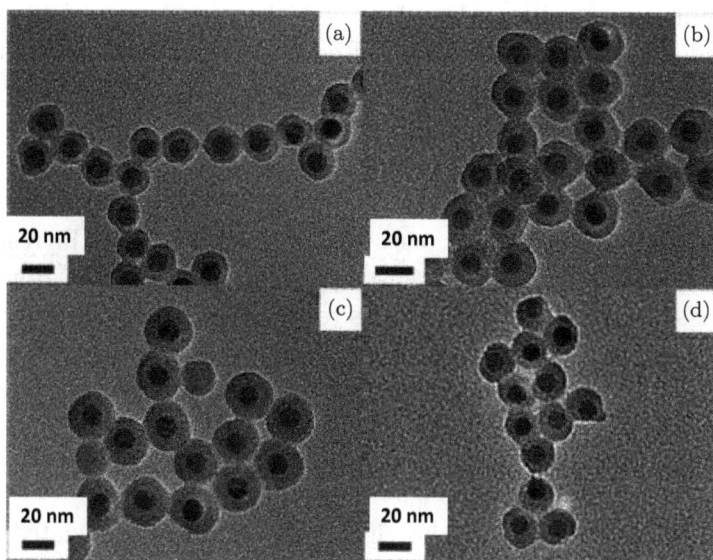

Figure 3.16. TEM images of Fe$_3$O$_4$@SiO$_2$ NPs of samples. (a) C1, (b) C2, (c) C3 and (d) C4 (adapted from Ref. [43]).

this problem, the TEOS content was decreased from 75 to 35 μL. On decreasing the TEOS content, the supply rate of monomers can be decreased which decreases the homogeneous nucleation of monomers.

3.7   Properties of SMNPs

3.7.1   Chemical properties

SMNPs possess a distinct structure with a high density of silanol groups on their surface which can be tailored by a broad range of inorganic and organic functional groups. Surface modification of SMNPs can offer various applications in nanomedicine. It also enhances various properties, such as amplification of host–guest interaction, control of charge density on the surface and grafting of various functional active molecules (prodrugs, MRI active moieties and fluorescent molecules).

Chang *et al.* described that surface modification of SMNPs can be tuned by suitable functionalizing linkers/agents (such as amino, carboxyl, methyl phosphonate, phenyl, sulfonate, sulfamide and sulfamates and carboxylate) which offer a very promising methodology for the fabrication of advance heterogeneous catalytic system [44].

3.7.2   Biocompatibility

Water dispersibility: Various synthetic methods of MNPs result in the formation of organic solvent-dispersible nanoparticles. However, their use in biomedical applications requires water dispersibility [45]. This property is administered by coating MNPs with silica precursor such as TEOS which is commonly employed to form a three-dimensional structure on the surface of magnetic core and functionalize the nanoparticles by providing OH terminal groups [46].

Colloidal stability: Another inevitable problem associated with bare MNPs in nanometer size range is their inherent instability, which leads to the formation of agglomerates [47]. Besides, naked MNPs are very much active and are simply oxidized with air, resulting in the

Table 3.6. Surface area of MNPs and SMNPs.

Material	S_{BET} (m^2/g)	Langmuir specific surface area (m^2/g)	Pore volume (cc/g)	Average pore diameter (nm)
MNPs	98.7	118.3	2.58	3.1
SMNPs	272.5	296.4	4.7	3.7

loss of magnetism and dispersibility. In this manner, silica coating acts as a protective coating and provides colloidal stability [48].

3.7.3 Physical properties

Surface area: In the domain of nanocatalyst, MNPs are largely employed in various applications due to their large surface area-to-volume ratio. Kalantary and co-workers described in their report that silica-coated MNPs provide larger surface area than bare MNPs (Table 3.6) [44, 49]. The enhancement of surface area of SMNPs can be attributed to special textural structure and high pore size area. This property of SMNPs is beneficial from both catalytic and adsorption capacity viewpoints.

3.7.4 Thermal properties

MNPs are highly vulnerable toward aerial oxidation at temperatures above 150°C [50]. On heating, in the presence of air, magnetite nanoparticles are converted to maghemite (γ-Fe$_2$O$_3$) and afterward to hematite (α-Fe$_2$O$_3$). Similarly, on laser irradiation, heating effect of light is sufficient to transform magnetite into other forms. SMNPs display outstanding stability at high temperature as compared to bare MNPs. Thermally stable silica shell protects MNPs from getting thermally oxidized. Due to this, SMNPs have been extensively studied in biomedical application, magnetic hyperthermia and heat emission blocking followed by exposure to magnetic field.

Setyawan and co-workers reported that SMNPs are thermally stable [51]. Figure 3.17 illustrates the thermal gravimetric analysis (TGA) curves of MNPs (S1) and SMNPs (S2). It can be seen that S1

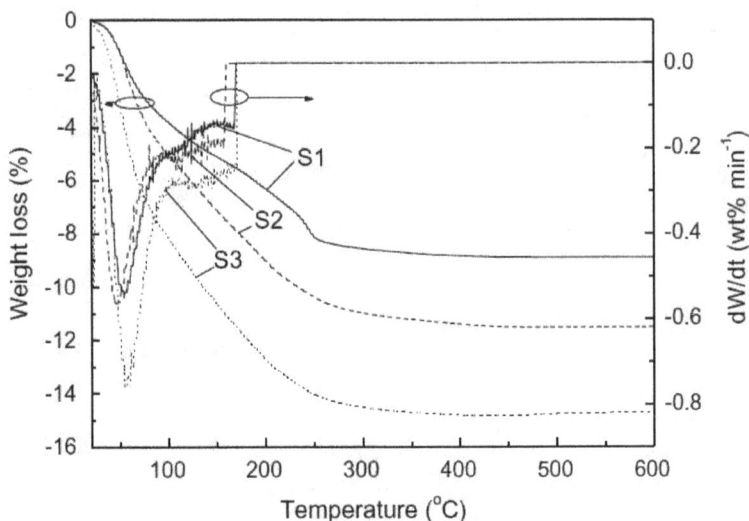

Figure 3.17. TGA curves of MNPs (S1) and SMNPs (S2) (adapted from Ref. [51]).

and S2 exhibit a weight reduction of 8.83% and 11.01%, respectively. Also, the shape of the two TGA curves is very comparable, which confirmed that SMNPs are thermally stable, and the slight difference in weight reduction occurs because of the physically adsorbed species on SMNPs other than silica.

3.7.5   Magnetic properties

Magnetic properties of SMNPs are generally studied using a vibrating sample magnetometer (VSM) at room temperature. Figure 3.18 displays curves showing decrease in the saturation magnetization (M_s) values from MNPs (58 emu/g) to SMNPs (36 emu/g). This decrease is due to the presence of non-magnetic silica. Although silica decreases the M_s value, SMNPs can be effortlessly removed from the medium *via* an external magnet.

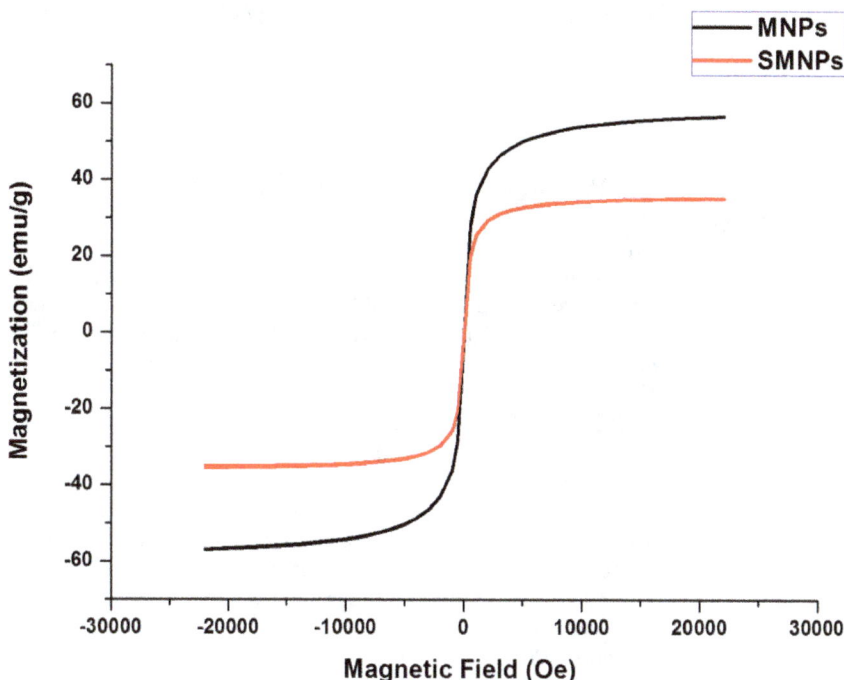

Figure 3.18. VSM curves of MNPs and SMNPs.

3.8 Conclusion

In this chapter, we have discussed the inevitable role of MNPs in catalysis. Further, benefits of coating silica on MNPs have been highlighted, such as prevention of magnetic interactions, easy functionalization, dispersion, improved thermal and chemical stability and biocompatibility. Silica coating can be done by various means, such as sol–gel, silicic acid, aerosol pyrolysis, arc discharge and microemulsion. Also, strategies are emphasized by which free-core silica or multicore silica nanoparticles (SNPs) can be avoided, resulting in the formation of continuous and homogeneously coated SMNPs. These particles possess fascinating surface, chemical, thermal and magnetic properties which make them an ideal candidate for catalytic support. However, these SMNPs must be surface-modified in order to be utilized for particular applications, which will be discussed in Chapter 4.

References

[1] (a) S. P. Gubin, Y. A. Koksharov, G. Khomutov, G. Y. Yurkov, *Russian Chemical Reviews* **2005**, *74*, 489–520; (b) T. Hyeon, *Chemical Communications* **2003**, 927–934; (c) A. H. Lu, E. L. Salabas, F. Schüth, *Angewandte Chemie International Edition* **2007**, *46*, 1222–1244; (d) J. Gao, H. Gu, B. Xu, *Accounts of Chemical Research* **2009**, *42*, 1097–1107; (e) S. Laurent, D. Forge, M. Port, A. Roch, C. Robic, L. Vander Elst, R. N. Muller, *Chemical Reviews* **2008**, *108*, 2064–2110.

[2] G. Reiss, A. Hütten, *Nature Materials* **2005**, *4*, 725–726.

[3] E. H. Kim, H. S. Lee, B. K. Kwak, B.-K. Kim, *Journal of Magnetism and Magnetic Materials* **2005**, *289*, 328–330.

[4] R. F. Ziolo, E. P. Giannelis, B. A. Weinstein, M. P. O'Horo, B. N. Ganguly, V. Mehrotra, M. W. Russell, D. R. Huffman, *Science* **1992**, *257*, 219–223.

[5] (a) F. Xu, J. H. Geiger, G. L. Baker, M. L. Bruening, *Langmuir* **2011**, *27*, 3106–3112; (b) C. Fang, M. Zhang, *Journal of Materials Chemistry* **2009**, *19*, 6258–6266; (c) M. Arruebo, R. Fernández-Pacheco, M. R. Ibarra, J. Santamaría, *Nano Today* **2007**, *2*, 22–32.

[6] (a) M. B. Gawande, Y. Monga, R. Zboril, R. Sharma, *Coordination Chemistry Reviews* **2015**, *288*, 118–143; (b) R. K. Sharma, S. Dutta, S. Sharma, R. Zboril, R. S. Varma, M. B. Gawande, *Green Chemistry* **2016**, *18*, 3184–3209.

[7] R. K. Sharma, A. Puri, Y. Monga, A. Adholeya, *Journal of Materials Chemistry A* **2014**, *2*, 12888–12898.

[8] G. Gnanaprakash, S. Mahadevan, T. Jayakumar, P. Kalyanasundaram, J. Philip, B. Raj, *Materials Chemistry and Physics* **2007**, *103*, 168–175.

[9] M. H. A. Latif, F. L. Yahya, *Chemistry and Materials Research* **2015**, *7*, 55–62.

[10] F. N. Sayed, V. Polshettiwar, *Scientific Reports* **2015**, *5*, 9733.

[11] X. Yang, C. Liu, J. Ahner, J. Yu, T. Klemmer, E. Johns, D. Weller, *Journal of Vacuum Science & Technology B: Microelectronics and Nanometer Structures Processing, Measurement, and Phenomena* **2004**, *22*, 31–34.

[12] J. E. Muñoz, J. Cervantes, R. Esparza, G. Rosas, *Journal of Nanoparticle Research* **2007**, *9*, 945–950.

[13] S. Wan, J. Huang, H. Yan, K. Liu, *Journal of Materials Chemistry* **2006**, *16*, 298–303.

[14] A. K. Ganguli, T. Ahmad, S. Vaidya, J. Ahmed, *Pure and Applied Chemistry* **2008**, *80*, 2451–2477.

[15] S. Sun, H. Zeng, D. B. Robinson, S. Raoux, P. M. Rice, S. X. Wang, G. Li, *Journal of the American Chemical Society* **2004**, *126*, 273–279.

[16] L. Maurizi, F. Bouyer, J. Paris, F. Demoisson, L. Saviot, N. Millot, *Chemical Communications* **2011**, *47*, 11706–11708.

[17] C. Feldmann, H. O. Jungk, *Angewandte Chemie International Edition* **2001**, *40*, 359–362.

[18] (a) R. Hong, N. O. Fischer, T. Emrick, V. M. Rotello, *Chemistry of Materials* **2005**, *17*, 4617–4621; (b) M. Kim, Y. Chen, Y. Liu, X. Peng, *Advanced Materials* **2005**, *17*, 1429–1432.

[19] L. Cabrera, S. Gutierrez, N. Menendez, M. Morales, P. Herrasti, *Electrochimica Acta* **2008**, *53*, 3436–3441.

[20] L.-H. Hung, A. P. Lee, *Journal of Medical and Biological Engineering* **2007**, *27*, 1–6.

[21] A. Bharde, D. Rautaray, V. Bansal, A. Ahmad, I. Sarkar, S. M. Yusuf, M. Sanyal, M. Sastry, *Small* **2006**, *2*, 135–141.

[22] W.-W. Wang, Y.-J. Zhu, M.-L. Ruan, *Journal of Nanoparticle Research* **2007**, *9*, 419–426.

[23] M. A. Malvindi, V. De Matteis, A. Galeone, V. Brunetti, G. C. Anyfantis, A. Athanassiou, R. Cingolani, P. P. Pompa, *PloS One* **2014**, *9*, e85835.

[24] S. Santra, R. Tapec, N. Theodoropoulou, J. Dobson, A. Hebard, W. Tan, *Langmuir* **2001**, *17*, 2900–2906.

[25] S. Uthaman, S. J. Lee, K. Cherukula, C.-S. Cho, I.-K. Park, *BioMed Research International* **2015**, *2015*, 959175.

[26] (a) S. M. Silva, R. Tavallaie, L. Sandiford, R. D. Tilley, J. J. Gooding, *Chemical Communications* **2016**, *52*, 7528–7540; (b) L. Wang, J. Luo, S. Shan, E. Crew, J. Yin, C.-J. Zhong, B. Wallek, S. S. Wong, *Analytical Chemistry* **2011**, *83*, 8688–8695; (c) Z. Xu, Y. Hou, S. Sun, *Journal of the American Chemical Society* **2007**, *129*, 8698–8699.

[27] (a) S. Dutta, S. Sharma, A. Sharma, R. K. Sharma, *ACS Omega* **2017**, *2*, 2778–2791; (b) R. Gaur, M. Yadav, R. Gupta, G. Arora, P. Rana, R. K. Sharma, *ChemistrySelect* **2018**, *3*, 2502–2508; (c) R. Gupta, M. Yadav, R. Gaur, G. Arora, R. K. Sharma, *Green Chemistry* **2017**, *19*, 3801–3812; (d) R. K. Sharma, S. Dutta, S. Sharma, *Dalton Transactions* **2015**, *44*, 1303–1316; (e) R. K. Sharma, Y. Monga, *Applied Catalysis A: General* **2013**, *454*, 1–10; (f) R. K. Sharma, Y. Monga, A. Puri, *Catalysis Communications* **2013**, *35*, 110–114; (g) R. K. Sharma, Y. Monga, A. Puri, *Journal of Molecular Catalysis A: Chemical* **2014**, *393*, 84–95; (h) R. K. Sharma, A. Puri, Y. Monga, A. Adholeya, *Separation and Purification Technology* **2014**, *127*, 121–130; (i) R. K. Sharma, S. Dutta, S. Sharma, *New Journal of Chemistry* **2016**, *40*, 2089–2101; (j) R. K. Sharma, R. Gaur, M. Yadav, A. Goswami, R. Zbořil, M. B. Gawande, *Scientific Reports* **2018**, *8*, 1901; (k) R. K. Sharma, Y. Monga, A. Puri, G. Gaba, *Green Chemistry* **2013**, *15*, 2800–2809; (l) R. K. Sharma, M. Yadav, R. Gaur, R. Gupta, A. Adholeya, M. B. Gawande, *ChemPlusChem* **2016**, *81*, 1312–1319; (m) R. K. Sharma, M. Yadav, R. Gaur, Y. Monga, A. Adholeya, *Catalysis Science & Technology* **2015**, *5*, 2728–2740; (n) R. K. Sharma, M. Yadav,

M. B. Gawande, *Ferrites and Ferrates: Chemistry and Applications in Sustainable Energy and Environmental Remediation*, ACS Publications, **2016**, pp. 1–38; (o) R. K. Sharma, M. Yadav, Y. Monga, R. Gaur, A. Adholeya, R. Zboril, R. S. Varma, M. B. Gawande, *ACS Sustainable Chemistry & Engineering* **2016**, *4*, 1123–1130.

[28] (a) L. T. Mai Hoa, T. T. Dung, T. M. Danh, N. H. Duc, D. M. Chien, *Journal of Physics: Conference Series* **2009**, *187*, 12048; (b) Q. M. Kainz, O. Reiser, *Accounts of Chemical Research* **2014**, *47*, 667–677; (c) F. Li, J. Sun, H. Zhu, X. Wen, C. Lin, D. Shi, *Colloids and Surfaces B: Biointerfaces* **2011**, *88*, 58–62.

[29] (a) S. Zhang, H. Niu, Z. Hu, Y. Cai, Y. Shi, *Journal of Chromatography A* **2010**, *1217*, 4757–4764; (b) W. Fan, W. Gao, C. Zhang, W. W. Tjiu, J. Pan, T. Liu, *Journal of Materials Chemistry* **2012**, *22*, 25108–25115; (c) G. Arora, M. Yadav, R. Gaur, R. Gupta, R. K. Sharma, *ChemistrySelect* **2017**, *2*, 10871–10879.

[30] (a) S. Anandan, G.-J. Lee, S.-H. Hsieh, M. Ashokkumar, J. J. Wu, *Industrial & Engineering Chemistry Research* **2011**, *50*, 7874–7881; (b) L. Sun, C. Zhang, L. Chen, J. Liu, H. Jin, H. Xu, L. Ding, *Analytica Chimica Acta* **2009**, *638*, 162–168; (c) L. De Matteis, L. Custardoy, R. Fernández-Pacheco, C. Magén, J. M. de la Fuente, C. Marquina, M. R. Ibarra, *Chemistry of Materials* **2012**, *24*, 451–456.

[31] C. Jin, Y. Wang, H. Wei, H. Tang, X. Liu, T. Lu, J. Wang, *Journal of Materials Chemistry A* **2014**, *2*, 11202–11208.

[32] P. Tartaj, M. del Puerto Morales, S. Veintemillas-Verdaguer, T. González-Carreño, C. J. Serna, *Journal of Physics D: Applied Physics* **2003**, *36*, R182–R197.

[33] (a) Q. A. Pankhurst, J. Connolly, S. Jones, J. Dobson, *Journal of Physics D: Applied Physics* **2003**, *36*, R167–R181; (b) Y.-H. Deng, C.-C. Wang, J.-H. Hu, W.-L. Yang, S.-K. Fu, *Colloids and Surfaces A: Physicochemical and Engineering Aspects* **2005**, *262*, 87–93.

[34] W. Wu, Q. He, C. Jiang, *Nanoscale Research Letters* **2008**, *3*, 397–415.

[35] X. Liu, J. Xing, Y. Guan, G. Shan, H. Liu, *Colloids and Surfaces A: Physicochemical and Engineering Aspects* **2004**, *238*, 127–131.

[36] (a) T. G. Carreño, A. Mifsud, C. Serna, J. Palacios, *Materials Chemistry and Physics* **1991**, *27*, 287–296; (b) P. Tartaj, T. Gonzalez-Carreno, C. J. Serna, *Advanced Materials* **2001**, *13*, 1620–1624.

[37] O. B. Miguel, M. Morales, C. Serna, S. Veintemillas-Verdaguer, *IEEE Transactions on Magnetics* **2002**, *38*, 2616–2618.

[38] O. Bomatí-Miguel, Y. Leconte, M. Morales, N. Herlin-Boime, S. Veintemillas-Verdaguer, *Journal of Magnetism and Magnetic Materials* **2005**, *290*, 272–275.

[39] R. Fernández-Pacheco, M. Arruebo, C. Marquina, R. Ibarra, J. Arbiol, J. Santamaría, *Nanotechnology* **2006**, *17*, 1188–1192.

[40] M. A. Malik, M. Y. Wani, M. A. Hashim, *Arabian Journal of Chemistry* **2012**, *5*, 397–417.

[41] D. C. Lee, F. V. Mikulec, J. M. Pelaez, B. Koo, B. A. Korgel, *The Journal of Physical Chemistry B* **2006**, *110*, 11160–11166.

[42] C. Vogt, M. S. Toprak, M. Muhammed, S. Laurent, J.-L. Bridot, R. N. Müller, *Journal of Nanoparticle Research* **2010**, *12*, 1137–1147.

[43] H. Ding, Y. Zhang, S. Wang, J. Xu, S. Xu, G. Li, *Chemistry of Materials* **2012**, *24*, 4572–4580.

[44] B. Chang, J. Guo, C. Liu, J. Qian, W. Yang, *Journal of Materials Chemistry* **2010**, *20*, 9941–9947.

[45] S. Munjal, N. Khare, C. Nehate, V. Koul, *Journal of Magnetism and Magnetic Materials* **2016**, *404*, 166–169.

[46] J. Li, Y. Hu, J. Yang, W. Sun, H. Cai, P. Wei, Y. Sun, G. Zhang, X. Shi, M. Shen, *Journal of Materials Chemistry B* **2015**, *3*, 5720–5730.

[47] A. Akbarzadeh, M. Samiei, S. Davaran, *Nanoscale Research Letters* **2012**, *7*, 144.

[48] W. Wu, Z. Wu, T. Yu, C. Jiang, W.-S. Kim, *Science and Technology of Advanced Materials* **2015**, *16*, 023501.

[49] E. K. Pasandideh, B. Kakavandi, S. Nasseri, A. H. Mahvi, R. Nabizadeh, A. Esrafili, R. R. Kalantary, *Journal of Environmental Health Science and Engineering* **2016**, *14*, 21.

[50] K. Cendrowski, P. Sikora, B. Zielinska, E. Horszczaruk, E. Mijowska, *Applied Surface Science* **2017**, *407*, 391–397.

[51] F. Fajaroh, H. Setyawan, A. Nur, I. W. Lenggoro, *Advanced Powder Technology* **2013**, *24*, 507–511.

Chapter 4

Different Approaches for Surface Modification

Manavi Yadav, Yukti Monga, Gunjan Arora
and Rakesh Kumar Sharma*
*rksharmagreenchem@hotmail.com

4.1 Introduction

With the rapid development of nanoscience and nanotechnology, nanostructured composites, nanoparticles (NPs) in particular, have received considerable attention from a broad range of disciplines [1]. Among the numerous organic–inorganic nanohybrids, silica nanospheres (SNSs) have been widely employed in heterogeneous supported catalysts, electronics, sensor industries and in several biomedical applications. This is attributed to their biocompatibility, low toxicity, scalable synthetic availability and ease with which their properties can be tuned for diverse applications, such as precise control of their particle size, shape, porosity and crystallinity [2].

Despite several advantages associated with SNSs, one of their shortcomings is the inconvenience and inefficiency encountered while recovering them, which hampers their utility in many applications [3]. To overcome these challenges, magnetic NPs (MNPs) emerged as

---

*Corresponding author.

a promising solution as they not only combine the unique characteristics of NPs but also possess superior properties, such as facile magnetic recovery, low toxicity and cost-effectiveness [4]. Unlike the cumbersome separation procedures, the magnetic approach prevents the use of auxiliary substances (solvents, filters, etc.), thereby making the process swift, clean, green and eco-friendly [5]. In fact, remarkable features of MNPs, such as superparamagnetism, low Curie temperature and high magnetic susceptibility, have led to the integration of MNPs in a myriad of commercial applications. Due to their strong magnetic properties and low toxicity, presently, they are extensively employed in magnetic fluids, data storage, catalysis and biomedical applications, including targeted drug delivery, bioseparation, detection of biological entities (such as protein, enzymes, cell, nucleic acids and virus), cellular therapy such as cell labeling, tissue repair and clinical diagnosis such as magnetic resonance imaging (MRI) and magnetic fluid hyperthermia (MFH) [6].

But, with several advantages, there are some associated disadvantages of using naked MNPs. Due to the large surface-to-volume ratio, MNPs possess high surface energies [4a]. As a consequence, these NPs tend to aggregate so as to minimize their surface energies. Besides, bare MNPs have a high chemical activity, and hence are easily oxidized in air, thereby resulting in loss of magnetism and dispersibility. Therefore, it is crucial to provide appropriate surface coating by developing effective protection strategies to restore the stability of MNPs. This is usually performed by coating the surface with organic molecules, including surfactants, polymers and biomolecules, or coating with an inorganic layer, such as silica, metal oxide or metal sulfide, and a metallic or non-metallic substance. Indeed, it has been observed that the protecting layer not only is important for stabilizing the MNPs by engineering the surface energy and interfacial properties such as wetting, adhesion and friction but can also be utilized for further functionalization [7].

In general, the non-magnetic protecting shell suppresses the magnetic bipolar interaction and prevents them from agglomerating. Among all, silica is considered to be the most prominent material for coating because it is chemically inert, and therefore does

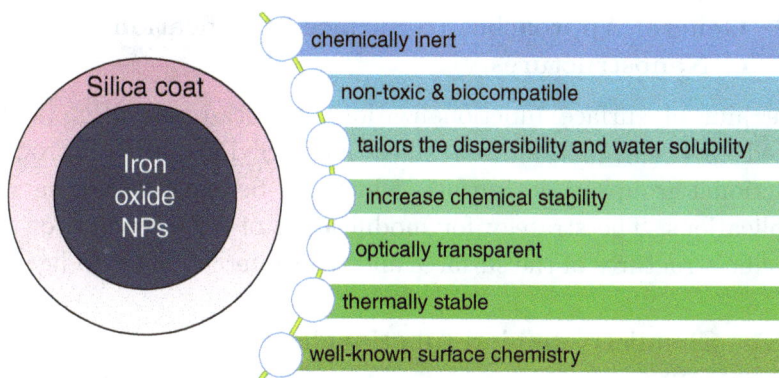

Figure 4.1. Benefits of silica as protective layer.

not affect the redox reaction at the core surface and maintains the physical characteristics of naked MNPs with high saturation magnetization [3]. Moreover, the surface of silica has terminal silanol groups which can be used to covalently attach further functionalities that can manipulate the characteristics of nanocomposites, thereby making them potential candidates for applications in catalysis, sensors, adsorbents and biomedical sciences [4b]. Figure 4.1 illustrates the benefits of silica coat on NPs.

With this, if one could combine the advantages of silica and magnetic core to fabricate a nanocomposite with high surface area and magnetic separability, a novel matrix can be developed that will have outstanding potential to stimulate extensive development in various applications. An essential requirement to explore the potential of these high-quality NPs is the fine-tuning and precise control of their surface chemistry. This is accomplished by surface modification. A range of surface modification approaches have been developed in order to improve the surface characteristics of NPs and to achieve explicit applications. This includes physical adsorption and covalent attachment of small molecules or polymers. In this chapter, we focus mainly on general approaches for surface functionalization of NPs and recent development in various strategies involved in modifying the surface of SNSs and silica-coated magnetic iron oxide NPs with their corresponding applications.

4.2 General Approaches to Surface Modification of Nanostructures

The aim of surface functionalization of silica-based NPs (silica-based MNPs (SMNPs) and silica NPs (SNPs)) is to introduce functional groups on its surface, which can be exploited for various applications. The strategy for modification of NPs depends on the specific structures of the surfaces and their interactions with ligands.

4.2.1 Chemistry behind functionalization

The silica surface is covered with hydroxyl groups that can be modified with organic or inorganic molecules, regardless of whether they are synthesized in aqueous or non-hydrolytic solutions. NPs which are synthesized using non-hydrolytic solutions need to be modified first to introduce functional groups such as hydroxyl and mercapto for further grafting reactions [8]. Amine and oxysilane are two other popular functional groups that are often used for surface modification of NPs [9]. The selected surface species not only prevent NPs from aggregation but also fulfill specific applications by grafting and activate the surface for further functionalization with other chemical or biological molecules [10]. Usually, these primers have dual functional groups, one for surface binding and the other for initiating the designed chemical reactions.

Silane agents such as 3-aminopropyltriethoxysilane (APTES) [11], *p*-aminophenyl trimethylsilane (APTS) [12] and mercaptopropyltriethoxysilane (MPTES) [13] are often considered as potential candidates for modifying the surface of MNPs directly. Such surface modification enhances the compatibility and provides rather high-density surface functional end groups which allow binding of other metals, polymers or biomolecules [14]. The existence of many hydroxyl groups on the surface of MNPs leads to the reaction between alkoxysilane groups and the formation of Si–O bond which supports the immobilization of the functional group. Thus, these silane coupling agents are compounds which can change the surface of NPs to be vitreophilic and make further deposition of coating layers plausible. Electrostatic interactions and other types of van der Waals interactions are the main driving forces behind the preliminary surface modification of NPs.

4.2.2 Strategies

In recent years, many functionalizing strategies have been developed and successfully employed. Emulsion and self-assembly are the two most commonly used strategies for the surface modification of SNPs with organic groups [15]. Whereas for surface modification with inorganic oxide layers, sol–gel methods are rather useful [16], the Stöber method is also the most widely used strategy [17].

4.2.2.1 *Adsorption and self-assembly*

Adsorption *via* physical means is among the methods that are conceptually straightforward in the surface modification of NPs for good stability and hydrophilicity in suspensions. The experimental conditions used are generally not complicated and can therefore be conducted under ambient conditions.

Self-assembly is a method capable of making one-, two- and three-dimensional structures of nanomaterials. While self-assembly of planar structures has been widely studied, it is not optimal when nanometer-sized colloidal particles are involved. The major driving forces for self-assembly include electrostatic interactions, surface tension, capillary forces, hydrophobic interactions and biospecific recognition [18]. Host–guest interactions are not only typically seen in biological systems but also regularly used in the assembly of non-biological molecules through these weak interactions. Long carbon chain surfactants are routinely used as stabilizing agents and serve as a good platform for host–guest interactions [19].

Silica has been widely used since the invention of the Stöber method originally designed for the preparation of SNPs with well-controlled spherical shape and size using alcoholic solvents, catalysts and alkoxide precursors. The hydrolysis and polycondensation of silica alkoxide precursors can be catalyzed by either base or acid.

4.2.2.2 *Sol–gel method*

Sols are small colloidal NPs in solution and can form an interconnected network of metal oxides, which are called gels, upon further polycondensation in the presence of acid or base catalysts [16]. The widely used silica precursors are short carbon chain alkoxyl silanes, particularly tetraethyl orthosilicate (TEOS). Ammonia and sodium hydroxide (NaOH) are the typical base catalysts, while hydrochloric acid (HCl) and nitric acid (HNO$_3$) can be used as acid catalysts for the hydrolysis and condensation of TEOS in water and alcohol mixtures [20].

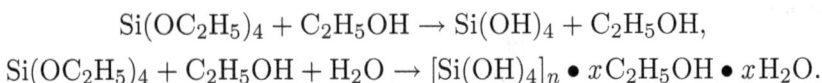

$$Si(OC_2H_5)_4 + C_2H_5OH \rightarrow Si(OH)_4 + C_2H_5OH,$$

$$Si(OC_2H_5)_4 + C_2H_5OH + H_2O \rightarrow [Si(OH)_4]_n \bullet x\,C_2H_5OH \bullet x\,H_2O.$$

With TEOS as the precursor, hydrolysis leads to the formation of silanol and ethanol in the solvent mixture during the first step. These silanol groups cross-link to form oligomers which further condense catalytically through –OH interactions on the surfaces of NPs and form silica networks. The second and third steps are the polycondensations where dense silica networks are formed. Other silane and halide precursors have also been studied for surface modification of NPs among which 3-aminopropyltrimethoxysilane (APTMS) and methacryloxypropyltrimethoxysilane (MPTMS) are the most widely used [21].

4.2.2.3 *Stöber method*

This method is mainly used for the surface modification of NPs with silica coatings. Since its invention, it has been modified and improved for use on NPs with different compositions, sizes, shapes and surface chemistry. The Stöber method has several advantages. First, the synthesis can take place in solvents with a wide range of hydrophilicity or hydrophobicity [22]. A primer can be introduced if necessary to activate the surfaces of NPs synthesized in highly hydrophobic solvents. Second, the formation of silica shells not only prevents the NPs from coalescing but also generates functional surfaces for further modification. Silica-coated NPs are often dispersed readily in aqueous solutions. Third, a silica coating can improve the biocompatibility of NPs due to its low toxicity. Therefore, the Stöber method is widely used in modifying spherical NPs, since achieving uniformity in coated layers for nanowires, nanocubes, nanorods and nanobars is feasible [23].

4.3 Prime Applications of Surface-Modified Silica-Based Nanomaterials

4.3.1 Biomedical applications

On account of the rich and well-documented biocompatibility of a silica coat, surface-modified SNPs/silica-coated MNPs (SMNPs) have been exploited for practical implementation in magnetically assisted drug delivery, tumor targeting and chemical separation of cells and proteins. There are reports where SMNPs have been employed to extract and purify genomic DNA and amino-modified SMNPs (ASMNPs) have been deployed to separate bacterial pathogens. Therefore, the effective introduction of an organic functional group on the surface of SMNPs has been the major object of investigation and research [24].

4.3.1.1 *Drug delivery*

For targeted drug delivery, silica-coated NPs with proper surface architecture and conjugated targeting ligands have attracted a great deal of attention. In order to increase the targeting yield, molecules such as carboxyl groups, biotin, avidin, carbodiimide and others can be functionalized on a silanol surface, and these further act as attachment points for the coupling of cytotoxic drugs or target antibodies to the carrier complex.

Recently, Chen *et al.* developed a multifunctional theranostic drug delivery system based on magnetic mesoporous silica NPs (MMSNs) and applied it for tumor-targeted MRI and precisely controlled therapy [25]. MMSNs were constructed using platinum (IV) prodrug, β-cyclodextrin and cancer-targeted peptide adamantane–PEG_8–glycine–arginine–glycine–aspartic–serine (AD–PEG_8–GRGDS) progressively. It was demonstrated that MMSNs

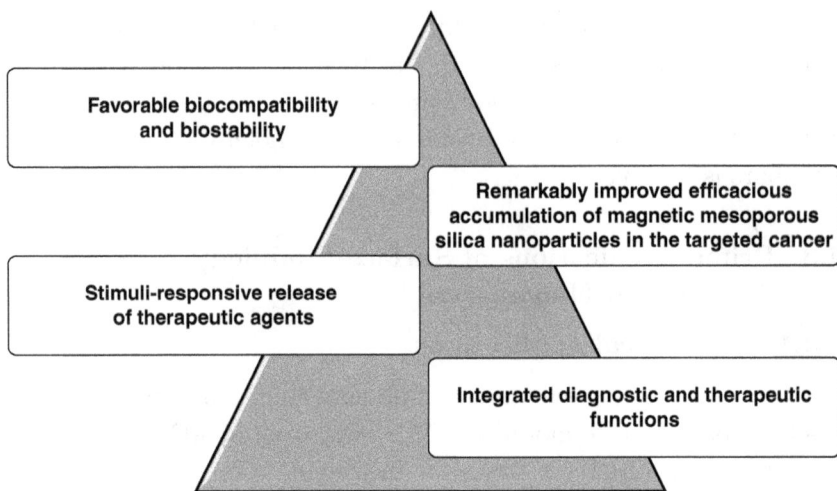

Figure 4.2.   Advantages of silica nanocomposites in drug delivery.

possessed competence of selective uptake of cancer cells and induced intracellular redox-sensitive release of anticancer drug to eradicate cancer cells efficiently. Furthermore, the synthesized nanocomposite also displayed high contrast in MRI for locating a tumor *in vivo* and achieved significant antitumor efficacy with minimum side effects. In the whole process, it was anticipated that incorporating a silica coat on a magnetic core could attain the advantage of silica, without sacrificing the unique magnetization characteristics. Some more advantages of silica coat are depicted in Figure 4.2.

For cancer therapies, thermal treatments such as hyperthermia and cryoablation are considered to have paramount importance [26]. But this requires changes in body temperature that can have potential damage. In this regard, NPs have fine control over heat localization, and MNPs have gained tremendous attention due to their inherent imaging contrast and ability to generate heat on exposure to alternating magnetic field. However, due to the problem of agglomeration of MNPs, both heating capacity and imaging performance get reduced under biological suspensions and are considered to impact *in vivo* biodistribution. Recently, Hurley *et al.* [26] have demonstrated a functionalized mesoporous silica shell that provides resistance to biological aggregation, protects the magnetic core from degradation and enables the practical use of high-heating, high-imaging capacity *in vitro* and *in vivo*. Moreover, it possesses large surface area and pore volume that can hold big molecular cargo and breaks down slowly in the body *via* slow dissolution. It was also found that this protective coating was minimally toxic toward neonatal human dermal fibroblasts.

Zhao and his group have fabricated biocompatible and multifunctional mesoporous silica NPs (MSNs) for *in vivo* cancer-targeted drug delivery. The authors synthesized poly(ethylene glycol)-incorporated MSNs with varied sizes and functionalized with amino-β-cyclodextrin bridged by cleavable disulfide bonds. Amino-β-cyclodextrin was used to trap drugs inside the mesopores. The activity of modified MSNs (48 nm) was demonstrated by incorporating an active folate-targeting ligand onto MSNs that exhibited substantial inhibition of tumor growth in mice treated with doxorubicin-loaded NPs. Loaded MSNs showed improved *in vivo* efficacy as compared to free doxorubicin and non-targeted NPs [27].

4.3.1.2 *Enzyme and biocide immobilization*

Enzyme immobilization on MNP is another process that creates a class of novel materials that possess the characteristics of both biological and heterogeneous catalysts. If an enzyme is grafted onto MNPs in an irreversible manner, it could lead to significant gain in catalytic properties of enzyme such as regenerability and reusability, thereby economizing the whole bioprocess. Recently, it has been established that SMNPs are non-toxic to the growth of living cells, with negligible inhibition effect [28]. A recent example of immobilization of cholesterol oxidase (COD, cholesterol: oxygen oxidoreductase, EC 1.1.3.6) onto amino silica-modified superparamagnetic NPs of maghemite was reported. In this, first maghemite NPs were prepared by co-precipitation procedure followed by silica coating produced from sodium silicate solution. Then, amino silane coupling agent was used to modify the surface with terminal amino groups which were activated using glutaraldehyde, a bifunctional cross-linker. Finally, a strategy to immobilize COD onto the synthesized nanomaterial was presented. This enzyme is the primary enzyme involved in the pathway of cholesterol degradation in a number of soil bacteria. This protein catalyzes the oxidation of cholesterol to 4-cholesten-3-one in the presence of molecular oxygen (O_2), and hydrogen peroxide (H_2O_2) is formed as a side product [29].

Another work was reported by Esmaeilnejad-Ahranjani and co-workers [30], where core–shell structured poly(acrylic acid) (PAA)-coated Fe_3O_4 cluster@SiO_2 nanocomposite particles were used as support materials for lipase immobilization (Scheme 4.1). The primary role of lipase is to break down dietary fats into smaller molecules such as glycerol and fatty acids. However, their practical usage in native form is quite narrow because of their short life span and problems in their recovery and reusability. These drawbacks were eliminated by their immobilization on amine-functionalized magnetic silica nanocomposite particles. It was also observed that the grafting of lipases onto the low- and high-molecular-weight PAA-coated particles resulted in high activity and stability under various physical and chemical environments (maximum activity was shown at 50°C and

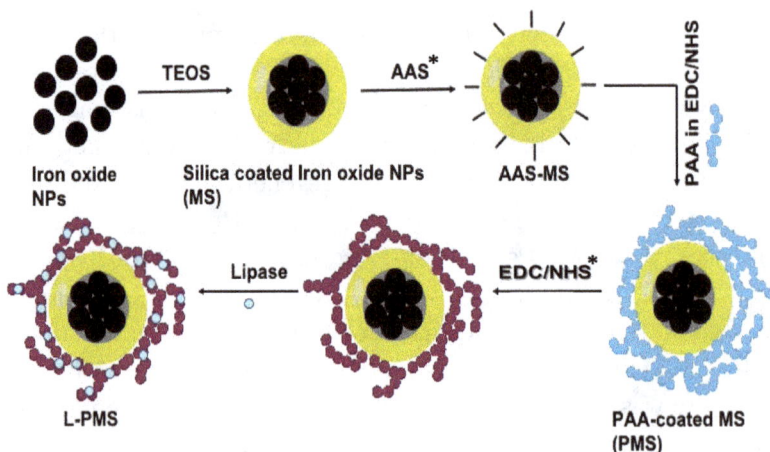

Scheme 4.1. Fabrication of core–shell structured PAA-coated Fe_3O_4 cluster@ SiO_2 nanocomposites as support materials for lipase immobilization.
Note: *AAS: 3-(2-aminoethyl amino)propyltrimethoxysilane, EDC: N-(3-dimethyl aminopropyl)-N-ethyl carbodiimide hydrochloride, NHS: N-hydroxysuccinimide.

55°C), in comparison to free lipase at its own optimum temperature (40°C). Moreover, the fabricated nanocomposite displayed marvelous performance at a broad range of pH and temperature conditions with superior reusability [30].

One more example was reported recently, where bare MNPs (both magnetite and maghemite), bimetallic gold (Fe_2O_3@Au) and SMNPs (Fe_3O_4@Si) were synthesized and further modified with 3-phosphonopropionic acid (3-PPA) to covalently immobilize α-amylase enzyme by using carbodiimide (EDC) [31].

Owing to the increasing resistance toward antibiotics, researchers are developing new strategies to combat infections caused by antibiotic-resistant bacteria. In this respect, a group recently grafted N-chloramine and quaternized N-chloramine onto magnetite NPs to generate antibacterial MNPs [32]. In this, two differently sized MNPs (3 and 10 nm) were fabricated and coated with silica and (3-chloropropyl)triethoxysilane (CPTES), followed by subsequent introduction of biocide N-chloramine precursors — dimethyl hydantoin (DMH) and quaternized dimethyl hydantoin (QDMH) (Scheme 4.2). These biocides are chemical species that destroy the microorganisms' growth and are considered better than antibiotics because of their multiple microbial targets and broader antimicrobial range. Among all, N-chloramines have garnered tremendous attention

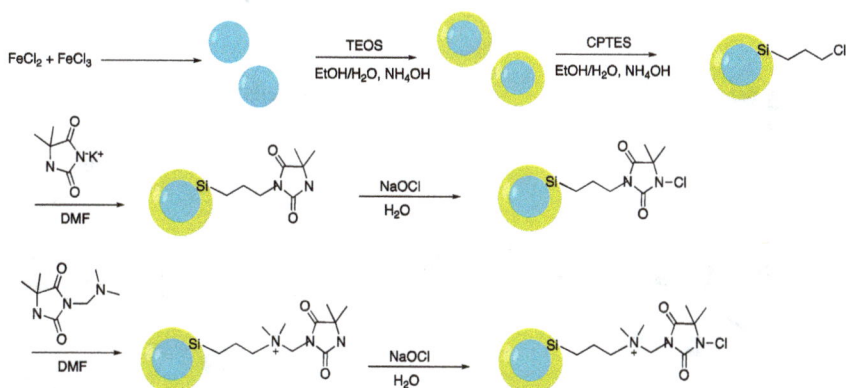

Scheme 4.2. Preparation of NPs with N-chlorohydantoin shell and Fe-oxide MNPs core.

because of their powerful antibacterial activity, long-term stability, reproducibility and an inherited lack of bacterial resistance. It was demonstrated that once the biocide is bound to the surface of NPs, it has a prolonged, localized and targeted effect on an infected wound. In this fashion, higher concentrations can be applied to target cells.

Liu *et al.* fabricated *N*-chloramine-functionalized hollow hemispherical silica particles as an effective antibacterial material. Hemispherical particles were synthesized by the sol–gel approach utilizing yeast cells as the template. This was followed by the fragmentation of the hollow particles by concentrated sulfuric acid and ultrasonication and, finally, the removal of cell cores was performed through bleaching (Scheme 4.3). It was hypothesized that the biocidal efficacy of the test material could be enhanced by increasing the contact surface between the antibacterial material and the bacteria. The antimicrobial property of these materials was explored in phosphate-buffered solutions and protein-enriched medium. The effect of chemical structures, surface charges and shape of the particles on their resistance to protein quenching was also investigated [33].

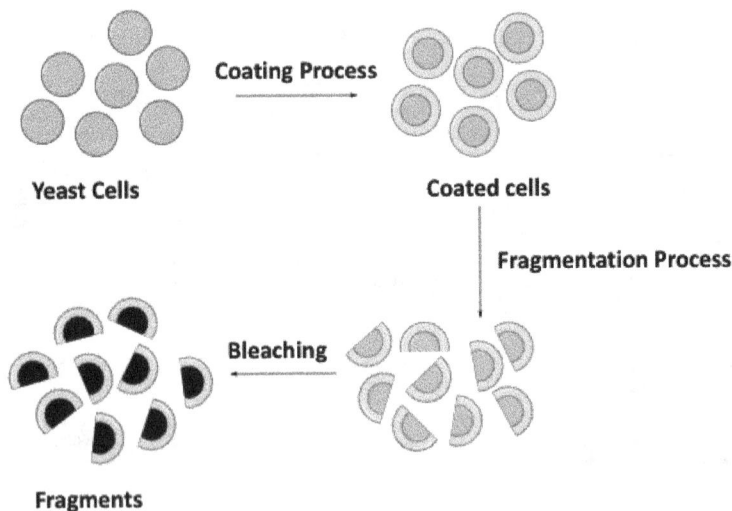

Scheme 4.3.   Preparation of SiO_2 and $CPTES/SiO_2@SiO_2$ fragments.

4.3.1.3 *Magnetic bioseparation*

Nowadays, magnetic bioseparation has become an interesting tool for biological molecules and cells because their magnetic properties enable facile separation of a target molecule from the complex mixture by an external magnetic field. For this, both magnetite (Fe_3O_4) and maghemite (Fe_2O_3) are considered as suitable cores due to their good biocompatibility, low toxicity and strong magnetization response [34].

In comparison to conventional bioseparation, magnetic separation has enormous advantages for detecting bacterial pathogens, viruses and transgenic crops. Functionalization of MNPs is considered to be the major element to capture target analytes efficiently. In this context, ASMNPs are considered to be a well-known tool for bioseparation as they can be readily modified by simple chemical reactions. Moreover, they can serve as capture agents through electrostatic interaction and hydrogen bonding between the amino group and some biological molecules, such as DNA. In recent times, one-pot synthesis of amino-rich SMNPs (ARSMNP) was reported, which was found to have a rough surface and high density of amine terminals than those prepared using post-modification technique. Also, the synthesized NPs were found to have good stability under acidic conditions and were deployed to absorb DNA directly. It was observed that the MNPs and DNA complexes could be directly used for the polymerase chain reaction (PCR) after magnetic separation and blocking without tedious work-up procedures. The whole process was found to display improved DNA capture efficiency for real-time quantitative PCR [35].

4.3.1.4 *Raman tags*

Raman spectroscopy is a useful technique for analyzing biological samples, but due to low intensities of these signals they cannot be directly used for bioassay. To overcome this, surface-enhanced Raman spectroscopy (SERS) appeared as a promising alternative as it increases the Raman intensities of molecules adsorbed at rough metal surfaces from 6 to 14 orders of magnitude [36]. However, there is still a need of labels in some cases to trace the interaction between the analyte and the reaction system. SERS is a sensitive detection analytical tool that is even better than fluorescence because of its much wider range of labeling.

With this, Gong and co-workers [89] recently developed a simple, sensitive and highly specific immunoassay for human fetoprotein (AFP), a tumor marker for the diagnosis of hepatocellular carcinoma. This was done by immobilizing immunoreagent components on the surface of silica-coated magnetic NPs. In this, biocompatible Ag/SiO_2 NP-based Raman tags modified with oligonucleotides were used as the detection probe, and the amino group-functionalized SMNPs with capture strands were employed as immobilization matrix and separation tool for the detection of HIV-related DNA sequences. It was observed that the use of MNPs as supports enhanced the surface area, provided low mass transfer resistance and helped in selective separation of immobilized enzyme by magnetic forces. Also, the silica coat not only improved stability but also aided in binding different biological ligands onto its surface. In comparison to the previously reported surface-enhanced Raman immunoassays, this strategy was found to be advantageous in two aspects: first, the high stability of Raman tags derived from the silica-coated silver core–shell nanomaterial and, second, the use of SMNPs as an immobilization matrix and separation tool, thus escaping tedious pretreatment and washing steps [29].

4.3.2   As adsorbents

Contamination due to heavy metals represents a great threat to the ecosystem. A huge amount of metallic waste is released into the environment from automobiles, combustion of coal and metal-lurgical industries, which enters into the food chain through various routes. Although high concentrations have a severe effect on living organisms, their gradual accumulation over the life span causes toxic effects [37]. Therefore, their removal is imperious, and several reliable methods have been developed to solve this, such as liquid–liquid extraction (LLE), precipitation/co-precipitation, cloud-point extraction and solid-phase extraction (SPE). Among them, SPE technique is considered to be the most superior one for the separation and preconcentration of metal ions from low concentration samples. This is because of their several merits over other techniques like less waste production, reduced consumption of organic solvents, reusability, reduced matrix effect, easy coupling with different detectors and eco-friendly nature [38]. However, lack of selectivity hampers their usage when coexisting metal species interfere with the metal ion of interest. Hence, development of adsorbents with increased selectivity and sensitivity has been a topic of research in scientific community for the past many years.

In addition to numerous biomedical applications of silica-coated materials, several protocols have been developed for the separation of metal ions from wastewater, soil and food samples. The direct determination and extraction of heavy metals from real samples is a challenging task due to their complexity and extremely low concentrations, usually below the detection limit of the available techniques [37].

In this regard, 3-mercaptopropyltrimethoxysilane-functionalized silica nano-hollow spheres (SNHS) were fabricated to effectively remove heavy metal ions (Hg(II), Pb(II) and Cd(II)) from the water samples (Scheme 4.4). The thiol-functionalized silica hollow nanostructures demonstrated a high metal loading capacity owing to the presence of densely immobilized thiol groups on its surface. Investigation of the adsorption kinetics and isotherm of heavy metal

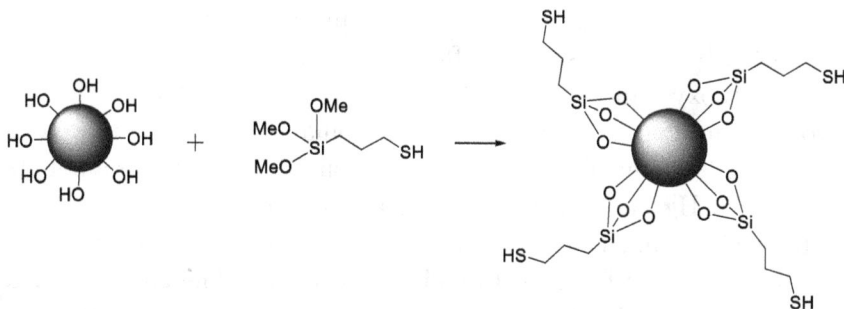

Scheme 4.4.   Synthesis of SNHS.

ions removal was performed in batch mode. The adsorption was found to depend on various factors, such as initial concentration of ions and contact time. The pseudo-second-order rate equation and Sips and Redlich–Peterson isotherms were applicable. Also, the percentage of removal is increasing, so thiol-functionalized adsorbent selectively removes ions from water, with high affinity for Hg (II), more than Pb(II) and Cd(II) [39].

Another report described the effective removal of Ni(II), Cd(II) and Pb(II) ions from water using amino-functionalized mesoporous and nano-mesoporous silica materials by Heidari *et al.* A series of experiments was performed to investigate the effects of the solution pH, metal ion concentrations, adsorbent dosages and contact time on metal adsorption. Improved adsorption of metal ions on the surface of the adsorbent was achieved with increasing solution pH and showed the maximum adsorption capacity of 12.36, 18.25 and 57.74 mg/g for Ni(II), Cd(II) and Pb(II), respectively [40].

Owing to the non-magnetic nature of dissolved contaminants, they do not respond to magnetic forces. Therefore, for the selective removal of toxic metals from environmental matrices, special functionalities with good affinities for target metal ions are grafted onto the surface of MNPs. In this respect, silica-coated MNPs have emerged as a sustainable solution because the silica shell consists of a stable and insoluble porous matrix that has suitable functional sites which can interact with the analytes [41]. Furthermore, it has very high thermal stability, does not swell, can achieve faster and quantitative sorption, has high adsorption capacity, high mass-exchange features, high reproducibility and large surface area [42].

Recently, SMNPs were functionalized with γ-mercaptopropyltrimethoxysilane (γ-MPTMS) in an efficient and cost-effective two-step method [8]. The resulting nanomaterials were deployed as an SPE adsorbent for separating and concentrating trace amounts of Cd, Cu, Hg and Pb from biological and environmental samples. Inductively coupled plasma-mass spectrometry (ICP-MS) was used to investigate the levels of the adsorbed metals. The effects of pH, sample volume, eluent and interfering ions were also investigated. It was observed that under the optimized conditions, the limits of

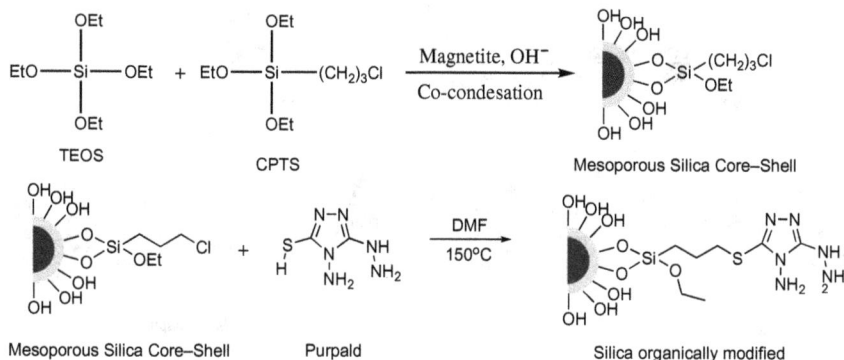

Scheme 4.5. Schematic representation of main reaction steps in sol–gel process for co-condensation method.

detection for Cd, Cu, Hg and Pb were as low as 24, 92, 107 and 56 pg L^{-1}, respectively.

Another work was reported where an ordered mesoporous Santa Barbara amorphous-type (SBA-15) silica with magnetic core was synthesized and its surface was modified with the ligand 4-amino-3-hydrazino-5-mercapto-1,2,4-triazole (Purpald), Scheme 4.5. The resulting nanoadsorbent was utilized to extract Cu(II) ions from aqueous medium through dispersive solid-phase microextraction (DSPME) and SPE column method [43].

For the adsorption of Cu^{2+} 4-{[(E)-phenylmethylidene]amino} benzoic acid, a Schiff base ligand was employed. The main strategy behind usage of this ligand was that it contains azomethine functional group that forms stable complexes with transition metals through its carbon–nitrogen double bond and can therefore be exploited for water purification. Ojemaye and co-workers [44] synthesized azomethine-functionalized MNPs (MNP-Maph) through the grafting of 4-{[(E)-phenylmethylidene]amino}benzoic acid (Maph-COOH) onto amino-functionalized MNPs (MNP-NH$_2$) by amide covalent coupling (Scheme 4.6). The performance of the resulting material was determined for the adsorption of Cu^{2+} from aqueous solution through batch adsorption studies, and to evaluate the optimum conditions for the effective removal of Cu^{2+} from polluted wastewater, the effects of various parameters were studied, such as pH of the

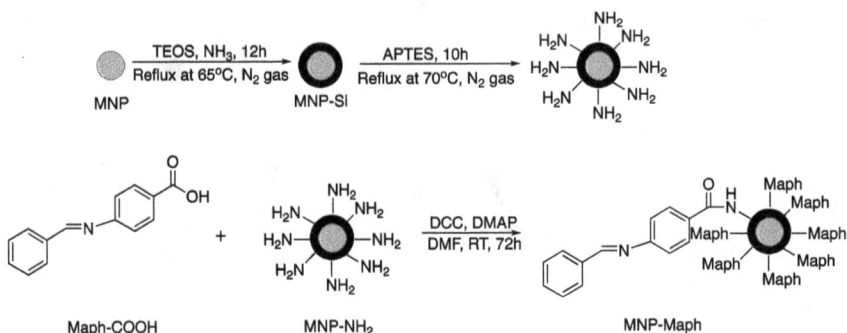

Scheme 4.6. Reaction pathway for the synthesis of azomethine-functionalized MNPs.

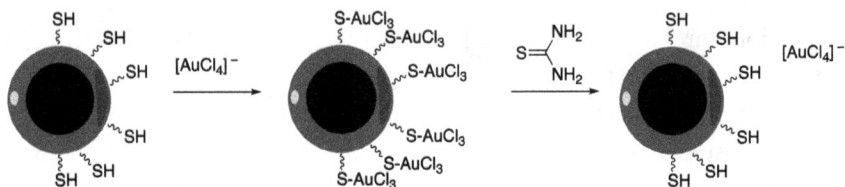

Scheme 4.7. Proposed interaction between $Fe_3O_4@SiO_2$ core–shell NPs and $[AuCl_4]^-$ ion.

solution, contact time, adsorbent dose, temperature and initial metal concentration [44].

Roto *et al.* fabricated a magnetic adsorbent of $Fe_3O_4@SiO_2$ core–shell NPs modified with thiol group for the adsorption of another trace metal ion chloroauric ($[AuCl_4]^-$), which is found to be corrosive to skin and a strong irritant to the eye and mucous membranes. For this, MNPs were prepared by co-precipitation method under mechanical stirring and coated with SiO_2 by acid hydrolysis of Na_2SiO_3 under N_2 purging. Further functionalization was performed using a thiol group, 3-mercaptopropyltrimethoxysilane, *via* silanization reaction. The concept of "hard and soft acids and bases (HSAB)" was exploited for extracting $[AuCl_4]^-$ as it is classified as a weak acid and thiol groups are classified as weak bases, thus providing specific interaction (Scheme 4.7) [45].

Another toxic metal ion, Cd^{2+} was adsorbed using chemically modified SMNPs. For this, SMNPs were functionalized with N-(2-aminoethyl)-3-aminopropyltrimethoxy-silane (EDS) to produce amine-functionalized magnetic nanocomposite particles (AF-MNPs). Onto this, a recombinant form of a rice metallothionein (MT) isoform was grafted to fabricate a novel nano-biohybrid magnetic adsorbent (MT-AF-MNPs) for highly effective removal of cadmium ions from aqueous media. It was confirmed by kinetic studies that adsorption of C on both AF-MNPs and MT-AF-MNPs is controlled by chemisorptions as both agreed well to pseudo-second-order kinetics. Also, it was assessed from Langmuir adsorption isotherm that adsorption by MT-AF-MNPs was considerably higher than that of the AF-MNPs due to high binding ability of MT toward Cd^{2+}. In addition to this, Cd-loaded adsorbents could be easily recovered and regenerated by acid treatment. The other advantages of this protocol include effortless magnetic separation, reusability and effective elimination of toxic Cd^{2+} from contaminated water [46].

Inspired by the exceptional behavior of MNPs, a class of thermosensitive and reusable MNPs were developed for treatment of emulsified oily wastewater. This was done by first preparing magnetic iron oxide by co-precipitation technique, followed by silica coating and surface functionalization with γ-methacryloxypropyl triisopropoxidesilane (MPS). To this, poly(N-isopropylacrylamide) (PNIPAM) molecular chains were attached *via* covalent grafting. To test the efficiency of synthesized nanoadsorbent, various parameters were evaluated, such as adsorbent dose, temperature, pH and reusability. It was observed that PNIPAM-grafted MNPs exhibited high demulsification efficiency at lower temperature, while at the lower critical solution temperature (LCST) of PNIPAM ($\sim$33°C), there was a dramatic decrease in its efficiency. It was also reported that these MNPs could be readily regenerated by rinsing with hot water and could be reused up to seven cycles without any considerable loss in its efficiency (Scheme 4.8) [47].

Among other toxic heavy metals known, lead is one of the most dangerous metals because of its carcinogenicity and high capacity to damage the central nervous system, brain, kidney,

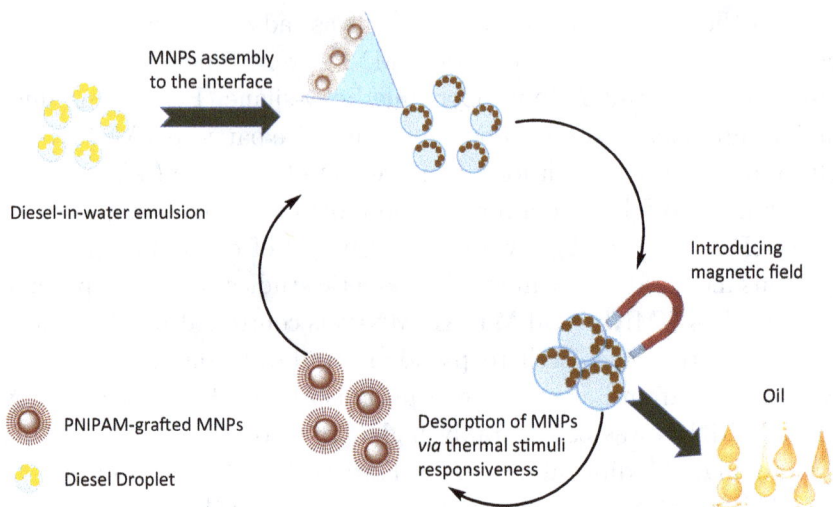

Scheme 4.8. Schematic process of recycling PNIPAM-grafted MNPs during oil–water separation.

and cardiovascular and reproductive systems [42]. Therefore, its removal at trace levels from environmental samples is one of the targets of analytical chemists. Recently, Culita and co-workers [48] presented a method to adsorb Pb(II) from aqueous solutions by synthesizing a novel magnetic adsorbent based on mesoporous silica-coated magnetite functionalized with Schiff base derived from *o*-Vanillin (Scheme 4.9). On account of popularity of *o*-Vanillin in coordination chemistry as it can readily produce variety of complexes because of its Schiff bases, it was employed for functionalizing the surface of SMNPs. The synthesized material was fully characterized by a variety of physicochemical techniques and then tested for Pb(II) uptake from aqueous solutions through the study of pH effect, contact time and initial concentration of Pb(II). The salient features of this protocol include chemical stability, improved adsorption capacity and facile recovery from reaction media by magnetic forces [48].

Knowing the hazardous nature of chromium ions, researchers developed ionic liquid-modified SMNPs (IL-SMNP) as efficient anion exchange sorbents for the extraction and determination of Cr (VI)

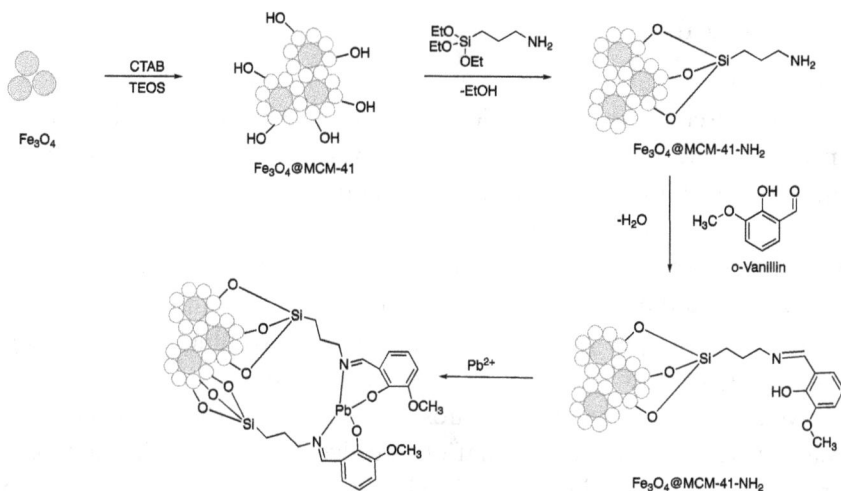

Scheme 4.9. Schematic illustration showing experimental process for magnetite functionalization and the subsequent adsorption of Pb(II) ions.

from tannery wastewater samples. It was also proposed that this work presents the tunability of IL-SMNPs, which can be used as both cation and anion exchange sorbents. The nanoadsorbent was characterized by SEM, FT-IR and VSM. A five-level rotatable central composite design (RCCD) was used for optimization of the extraction procedure. It was found that this method provided suitable analytical characteristics including low limit of detection (LOD), good repeatability and recovery. The other benefits of this protocol include its simplicity and time-saving extraction and desorption procedures in addition to its magnetic recoverability and good dispersibility [49].

Another work was reported where MNPs were synthesized to remove metals from binary solution. In this, diethylenetriamine-pentaacetic acid (DTPA)-functionalized SMNPs were utilized to adsorb Pb and Zn from single and bi-metallic metal solutions and from solutions containing dissolved organic carbon. It was also demonstrated that DTPA-functionalized NPs had the potential to be used as a remedial technology for both metal-contaminated water and solutions generated by soil washing [50].

4.3.3   Catalysis

Utilization of nanoscience has become an exponentially important research field in the development of environmental science, medicine and, most importantly, catalysis. The unique consistency and compatibility between Green Chemistry and nanoscience helps in solving environmental issues [51]. One of the 12 principles of Green Chemistry is the adoption of catalysts to achieve the goal of cleaner ecosystem, and an important class of nanoscience is nanocatalysts. A nano-sized catalyst possesses large surface-to-volume ratio, thereby increasing the rate of reaction due to increase of contact between the active site of catalyst and the substrate. However, their recovery and reuse after organic transformation are deciding factors in view of a green and sustainable protocol. Therefore, immobilization of these catalysts onto solid supports has received immense attention during quest of expansion of nanocatalysts [20].

Sharma and co-workers designed an SNS-supported copper catalytic system (SiO$_2$@APTES@DAP-Cu) by grafting diacetyl pyridine over amine-functionalized SNSs. The catalytic activity of the catalyst was investigated in oxidative amidation of methyl ketones using hydrogen peroxide as the oxidant and iodine as an additive. High efficiency, excellent functional group tolerance, milder reaction conditions, good recyclability and reusability are some of the advantages of the reported protocol [52].

The same group also fabricated an SNS-supported palladium catalyst (SiO$_2$@APTES@Pd-FFR) by immobilizing palladium complex onto amine-functionalized SNSs (Scheme 4.10). This was further employed as a catalyst in oxidative amination of aldehydes to commercially significant amides using H$_2$O$_2$ as a green oxidant. Features including mild reaction conditions, easy work-up, good yield, lack of need of toxic organic solvents, high turn-over frequency, facile recovery and reusability permit its use as an attractive alternative to the traditional catalytic methods for oxidative amination of aldehydes [53].

The choice of MNPs as solid support for the preparation of heterogeneous nanocatalysts appears to be an excellent sustainable

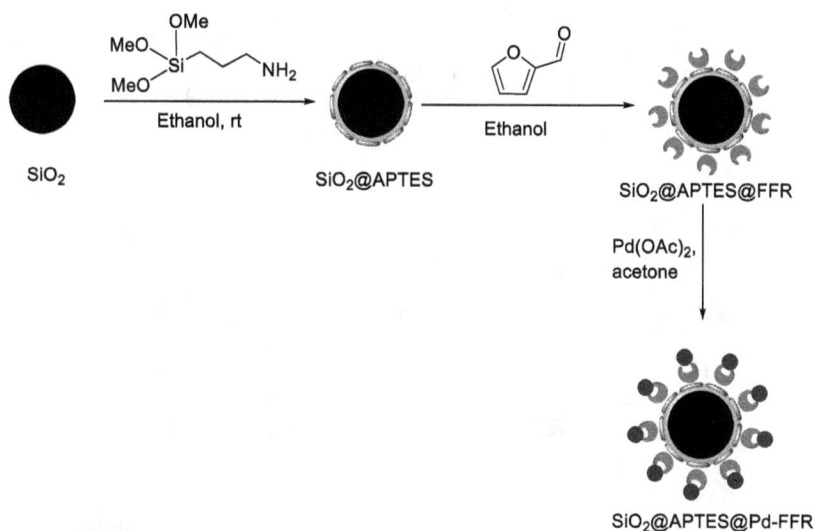

Scheme 4.10. Synthesis of SNS-based palladium nanocatalyst (SiO$_2$@APTES @Pd-FFR).

alternative to conventional materials as it covers almost all concerns on heterogeneous systems, such as activity, selectivity, work-up and recyclability. Recently, 4-(dimethylamino)pyridine (DMAP), an important organocatalyst, was grafted on MNPs *via* a new pathway through a ring opening of oxiran group (Scheme 4.11). This nanocatalyst (MNP-DMAP) was applied in a multicomponent reaction (MCR) for one-pot synthesis of a class of 2-amino-4H-chromene-3-carbonitrile derivatives having biological and pharmacological applications such as an anticoagulant, insecticidal and anticancer, antimicrobial, antibacterial and antiviral agents [54]. It was found to be an efficient heterogeneous catalyst for a three-component coupling reaction of aldehydes, beta diketones and malononitrile in one-pot synthesis of 2-amino-4H-chromene-3-carbonitrile under mild and green conditions. The as-synthesized catalyst exhibited reusability up to 10 times without significant loss in its activity.

In view of the importance of cross-coupling reactions, a nanocatalyst based on palladium supported on 3,3′-bisindolyl (4-hydroxyphenyl)methane-functionalized magnetite NPs was fabricated and applied for Sonogashira–Hagihara reaction [55]. Usually,

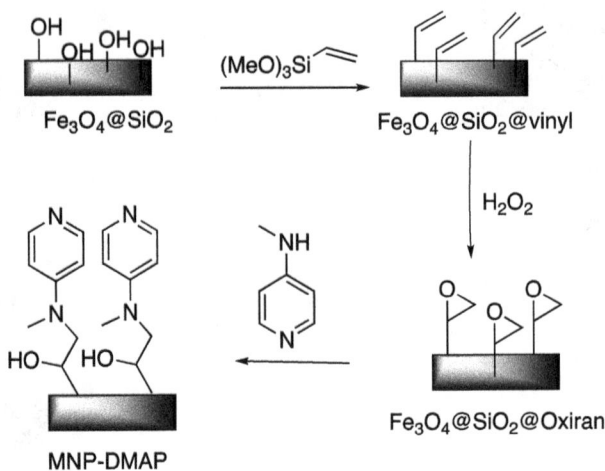

Scheme 4.11. Synthetic route for preparing MNP-DMAP catalyst.

this reaction is carried out in the presence of copper as a co-catalyst since copper enhances the reactivity of acetylene by the formation of copper acetylide. However, copper results in undesired homocoupling. Besides, phosphine ligands are used for this type of reaction which are considered to be toxic. This work represents Sonogashira–Hagihara reaction under copper-free and phosphine-free conditions. Also, alkynylation of a variety of aryl iodides and aryl bromides with terminal alkynes was carried out at 60°C using N,N-dimethyl acetamide as solvent, DABCO as base and low Pd loadings (0.18 mol%) under air. While for aryl chlorides, the reaction was carried out at 120°C in the presence of tetra-n-butylammonium bromide (TBAB) and 0.36 mol% of Pd catalyst, leaching test and hot-filtration tests were also conducted to confirm the heterogeneous nature of the catalyst. Easy recoverability and reusability up to seven consecutive runs without any appreciable loss of its catalytic activity are some additional features of this protocol.

Another class of heterogeneous catalyst was developed in recent times with a high loading of silver NPs onto the surface of silica core–shell MNPs coated with polymer (4-vinylpyridine) (Scheme 4.12). The developed catalyst ($Fe_3O_4@SiO_2–Ag$) was

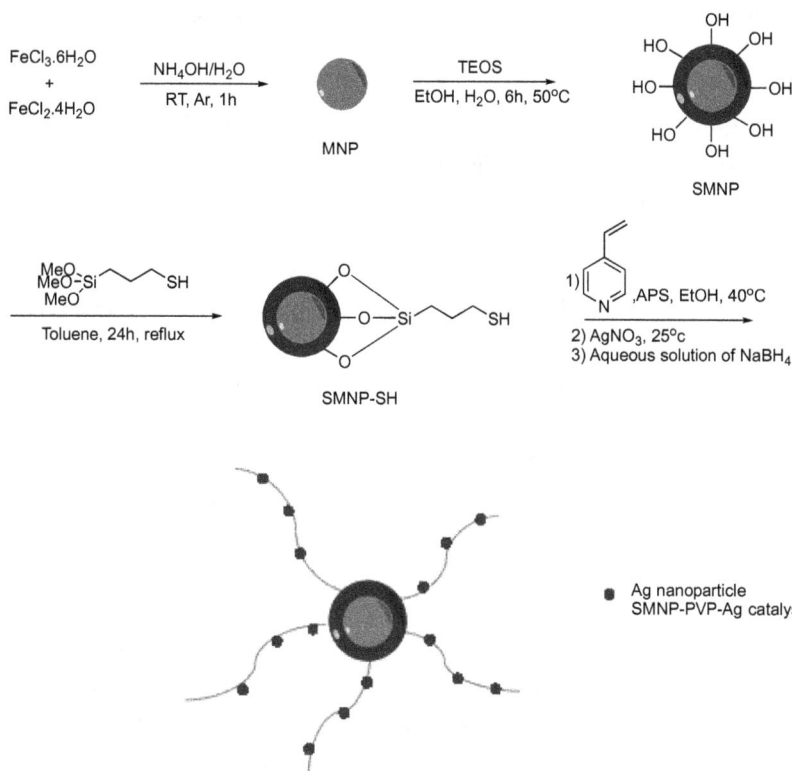

Scheme 4.12.   Synthesis of SMNP-PVP-Ag catalyst.

Note: PVP: poly(4-vinylpyridine); APS: ammonium persulfate.

employed in the dehydrogenation of alcohols to the corresponding carbonyl compounds. The role of this polymer coat was to enhance basicity and easily eliminate hydrogen atoms from alcohols, in addition to stabilizing and increasing the loading of Ag NPs. The catalyst was found to display excellent conversion for a broad diversity of alcohols into their corresponding carbonyl compounds. The catalyst was fully characterized by various physicochemical techniques. A vast range of aliphatic, cyclic, aromatic, heteroaromatic and benzylic alcohols with various steric and electronic effects were converted into their corresponding carbonyl compounds in good to excellent yields. The salient features of this protocol include simple synthesis, effortless magnetic recovery, reusability up to seven runs and high loading of Ag NPs [56].

Swami *et al.* presented a sulfonic acid-functionalized silica-coated $CuFe_2O_4$ magnetic nanocatalyst ($CuFe_2O_4@SiO_2–SO_3H$) and deployed it for synthesizing biologically and industrially relevant 2-pyrazole-3-amino-imidazo[1,2-a]-pyridine derivatives and related compounds *via* one-pot chemical synthesis of 2-aminoazine, alkyl-4-formyl-1-substituted phenyl-1H-pyrazole-3-carboxylate and iso-cyanides in ethanol. The synthesized nanocatalyst was characterized by XRD, TEM, SEM, FT-IR, EDX and ion exchange pH analysis. The strategy provided a superior alternative in comparison to the literature with several assets like high yield, operational simplicity, shorter reaction time, environmentally benign conditions and good tolerance to functional group. Moreover, the catalysts were found to be non-toxic, highly thermally and chemically stable, reusable and easily recoverable by applying external magnet to prevent unavoidable loss of NPs [57].

A possible mechanism was proposed for the synthesis of imidazo[1,2-a]pyridine derivatives and related compounds as shown in Scheme 4.13. First, the sulfonic group present on the outer surface of the nanocatalyst forms a hydrogen bond with the aldehyde oxygen that activates the carbonyl group to form carbonium ion. This facilitates the nucleophilic attack by aminoazine leading to the

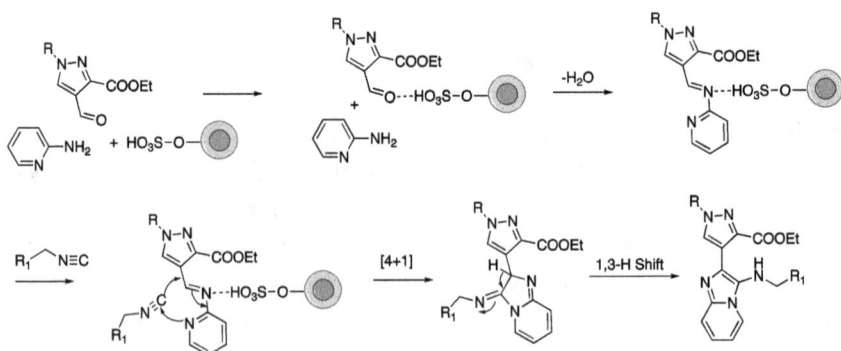

Scheme 4.13. Proposed mechanism for the synthesis of imidazo[1,2-a]pyridine derivatives.

formation of imine, and the resulting imine gets further activated by $CuFe_2O_4@SiO_2-SO_3H$ NPs to form electrophilic imine carbon which is subsequently attacked by isocyanide followed by [4 + 1] cycloaddition to give a cyclic adduct which finally undergoes 1,3-H shift to yield the desired product.

Similarly, a mild and green catalyst consisting of thiourea dioxide functionalized with chlorosulfonic acid on the surface of silica-coated MNPs $\{Fe_3O_4@SiO_2@(CH_2)_3$-thiourea dioxide-$SO_3H/HCl\}$ was developed. This was applied in the condensation of crotonaldehyde with indole at 60°C under solvent-free conditions to form the corresponding product. The attractive features of this work include its eco-friendly nature, easy purification, highly efficient, shorter reaction time and easy work-up [58].

Recently, Sharma *et al.* reported an efficient manganese-based magnetic nanocatalyst *via* covalent grafting of the manganese acetylacetonate complex onto amine-functionalized silica-coated iron-core NPs (Scheme 4.14). The overall process turned out to be greener as

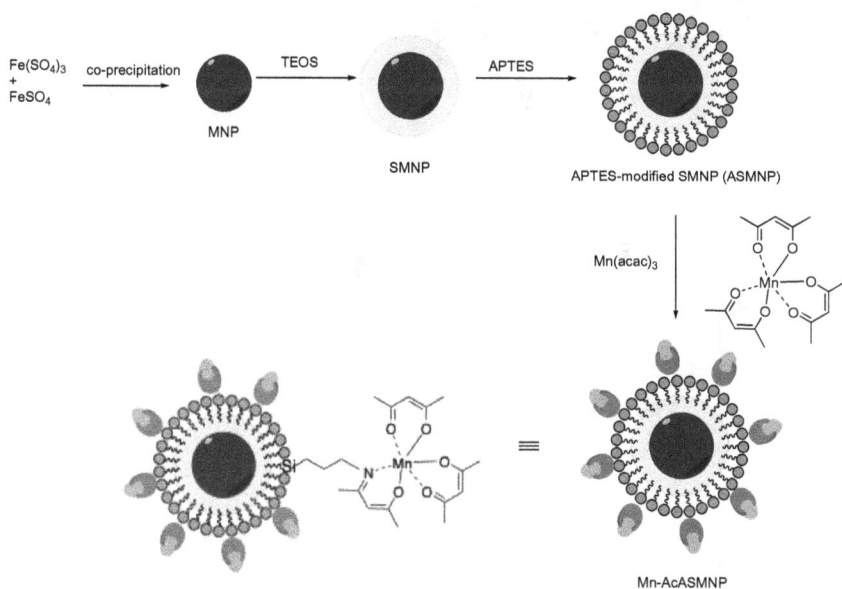

Scheme 4.14. Schematic illustration for the fabrication of Mn-Ac@ASMNP nanocatalyst.

Table 4.1. Common modifying strategies for functionalization of silica-coated NPs.

Modifying agent	Conditions	Structure of nanocomposite	Characterization	Ref.
Sulfuric acid	(i) SMNP, n-hexane, ultrasonic bath, 20 min (ii) Chlorosulfuric acid, 4 h, room temperature		FT-IR spectroscopy: Peaks at 640, 1220, 1230 and 2800–3700 cm^{-1} indicate sulfonic acid groups	[61]
Sulfuric acid	SNP, DCM, chlorosulfonic acid, 30 min		FT-IR spectroscopy: Band of the S–O vibrations appears at around 675 cm^{-1}, peaks at 1173 and 1286 cm^{-1} indicate symmetric and asymmetric S=O stretching	[62]

3-Phosphono propionic acid (3-PPA)	SMNP, 3-PPA, ultrasonication, 30 min		FT-IR spectroscopy: Peak at 1100–1150 min^{-1} corresponds to P–O and P–O–Fe stretching	[31]
Sulfamic acid	(i) SMNP, 3-chloropropyltrimethoxysilane, reflux, 24 h (ii) CH$_3$CN, ethylene diamine, triethylamine, reflux, 24 h		FT-IR spectroscopy: Appearance of the S=O stretching vibration bands at 1207 and 1127 cm^{-1}	[63]
Salicylic acid	Salicylic acid solution, acetic acid solution, 60°C, 2 h, nitrogen atmosphere		FT-IR spectroscopy: Bands at 1050 and 1150 cm^{-1} correspond to Si–OH, Si–O–Si vibrations and shifting correspond to immobilization of salicylic acid onto SMNPs	[64]

(Continued)

Table 4.1. (*Continued*)

Modifying agent	Conditions	Structure of nanocomposite	Characterization	Ref.
Sulfonic and ionic liquid groups bi-functionalized	(i) Brij-97:TEOS : MPTMS:CPTES = 433: 0.293:1:0.009:0.009, stir, 24 h, rt (ii) Resultant (MP + CP)-MSN:H$_2$O:MeOH: H$_2$O$_2$ = 0.5 g:10 mL: 10 mL:10 mL, rt, overnight (iii) (HSO$_3$ + CP)-MSN), imidazole, anhydrous benzene, chlorobutane, reflux (iv) Resultant compound dispersed in CrCl$_2$, overnight		NMR spectroscopy	[65]
Sulfonic acid	SNP, DCM, chlorosulfonic acid, stir		FT-IR spectroscopy: Peak at 1070 cm^{-1} indicates –SO$_3$ group	[66]

Folic acid	TEOS, hexadecyltrimethy-lammonium bromide (CTAB), reflux in acidic methanol	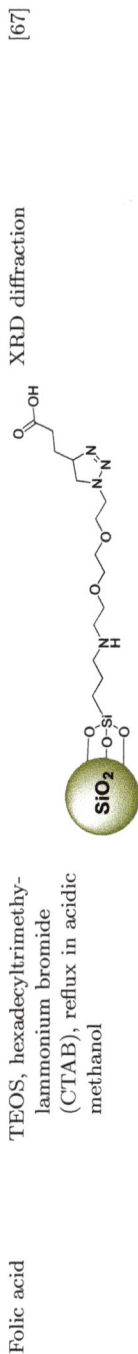	XRD diffraction	[67]
Diethylenetria minepentaacetic acid (DTPA)	(i) SMNP, (3-aminopropyl) triethoxysilane, overnight (ii) N,N-Dimethylformamide, triethylamine, DTPA di-anhydride, 80°C, 30 min (iii) Room temperature, 18 h	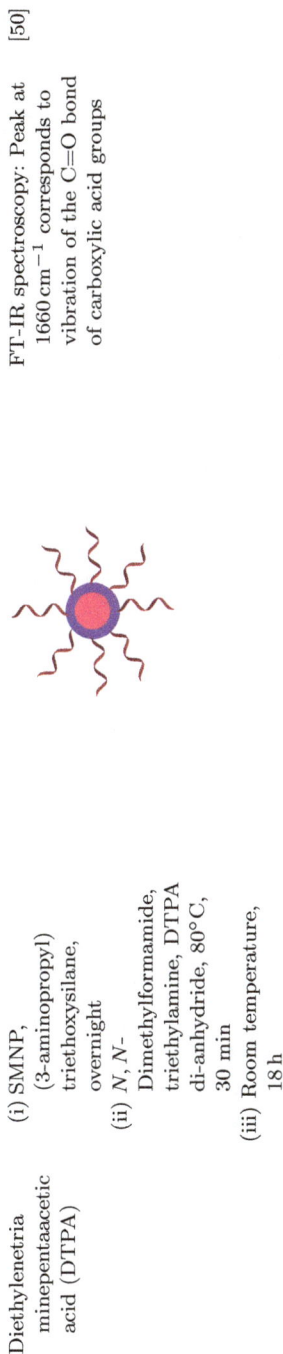	FT-IR spectroscopy: Peak at $1660\,cm^{-1}$ corresponds to vibration of the C=O bond of carboxylic acid groups	[50]
Polyethylenimine (PEI)	SMNP, aqueous solution of PEI, 1 M NaCl, overnight	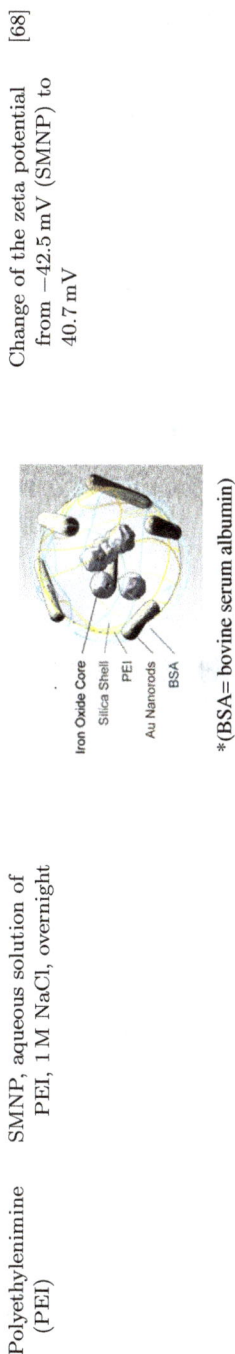 Iron Oxide Core Silica Shell PEI Au Nanorods BSA *(BSA= bovine serum albumin)	Change of the zeta potential from $-42.5\,mV$ (SMNP) to $40.7\,mV$	[68]

(*Continued*)

Table 4.1. (*Continued*)

Modifying agent	Conditions	Structure of nanocomposite	Characterization	Ref.
Polyaniline	(i) SMNP, distilled water, FeCl$_3$, 12 h (ii) Chloroform, aniline (60 μL), polymerized for 6 h		FT-IR spectroscopy: Bands at 1590 and 1495 cm^{-1} correspond to C–N and C–C stretching vibrations from the quinoid and benzenoid structure of PANI, respectively	[69]
Cysteine	(i) SMNP, dry toluene, 3-(triethoxysilyl) propyl isocyanate (ICPTES), reflux, 24 h (ii) Cysteine, THF, reflux, 5 h		FT-IR spectroscopy: Bands at 1635 cm^{-1} and 1570 cm^{-1} correspond to the formation of the urea [NHC(=O)NH] group and cysteine grafting, weak band near 2560 cm^{-1} indicates S–H group, the bands between 1200 and 1620 cm^{-1} correspond to the stretching and bending vibrations of C–O, C–N groups, and the broad band centered at 2980 cm^{-1} is assigned to asymmetric and symmetric stretching vibrations of –CH$_2$ groups	[70]

Polythiophene

Fe$_3$O$_4$@SiO$_2$, methanol, 3-vinylthiophene, AIBN (2,2′-Azobisisobuty-ronitrile), 60°C, 24 h

FT-IR spectroscopy: Peaks at 3040 cm^{-1} (CH, aromatic), 2980 cm^{-1} (CH, aliphatic) and 1560 cm^{-1} (C=C, aromatic)

[71]

Rodium–silica polymer brushes

(i) Si–PI, toluene, 4-diphenylphosphine styrene, polymerization in UV light (366 nm), (ii) Obtained polymer, RhCl$_3 \cdot x$H$_2$O, stir

^{31}P CP-MAS NMR spectroscopy: Strong signal at d value of 9 ppm, ^{29}Si CP-MAS NMR: Three signals at $d = 91, 101$ and 110 ppm

[72]

(3-Chloropropyl) trimethoxysilane

SMNP, dry toluene, (3-chloropropyl)trime-thoxysilane, reflux, 24 h, argon atmosphere

= Pd NPs

TEM analysis

[55]

(*Continued*)

Table 4.1. (*Continued*)

Modifying agent	Conditions	Structure of nanocomposite	Characterization	Ref.
Aminopropyltri ethoxysilane	SMNP, methanol, water, glycerol, aminopropyltriethoxysilane, 80–90°C, 3 h		FT-IR spectroscopy: Band at 2800–2900 cm^{-1} corresponds to C–H stretching vibrations due to aminopropyl chain	[73]
(3-Aminopropyl) trimethoxysilane	SMNP, aqueous ethanol (20%), (3-aminopropyl) trimethoxysilane, several hours		FT-IR spectroscopy: Band at 2924 cm^{-1} is related to the C–H stretching vibrations of anchored propyl chain	[74]
(3-Mercaptopropyl) trimethoxysilane	SMNP, dry toluene, (3-mercaptopropyl) trimethoxysilane, reflux, 24 h	= Ag NPs	FT-IR spectroscopy: Bands at 2850 and 2927 cm^{-1} correspond to C–H stretching and at 1652 and 3445 cm^{-1}	[56]

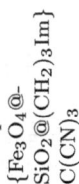

Fluoroalkyl silane (FAS)	PVDF-HFP and FAS in DMF, 30 min		FT-IR-ATR spectroscopy: Peaks at 1200 cm^{-1} correspond to C–F stretching vibration, peaks at 1710, 1250 and 1120 cm^{-1} correspond to the carbonyl stretching band, CH$_3$ and CH$_2$ of the underlying polyester fabric, respectively [75]
Ionic liquid {Fe$_3$O$_4$@-SiO$_2$@(CH$_2$)$_3$Im} C(CN)$_3$	(i) Toluene, 3-chloropropyltri-methoxysilane, triethylamine, reflux (ii) Toluene, imidazole, reflux (iii) CH(CN)$_3$, toluene, reflux		FT-IR spectroscopy: Peaks at 2947 and 2843 cm^{-1} represent the stretching vibrations of CH$_3$ and CH$_2$ groups, respectively [76]

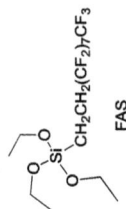

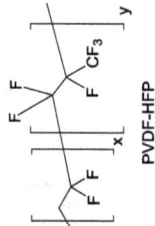

(Continued)

Table 4.1. (*Continued*)

Modifying agent	Conditions	Structure of nanocomposite	Characterization	Ref.
Ionic liquid methylimidazolium hexafluorophosphate (MIM-PF$_6$)	(i) SMNP, CPTMS, triethylamine, toluene, reflux, 48 h (ii) Toluene, N-methylimidazole, reflux, 48 h (iii) 7% KPF$_6$ aqueous solution		FT-IR spectroscopy: Sharp peak at 1,623 cm^{-1} corresponds to N-methylimidazolium ring	[77]
Ionic liquid (Triethoxysilyl-propylpyridinium hexafluorophosphate)	SMNP, ionic liquid, toluene, reflux, 24 h		FT-IR spectroscopy: Peaks at 2947 cm^{-1} and 2843 cm^{-1} represent the stretching vibrations of CH$_3$ and CH$_2$ groups, respectively, and at 1463 and 1381 cm^{-1} correspond to respective bending vibrations	[78]

Ionic liquid	(i) SMNP, (3-chloropropyl) triethoxysilane, dry toluene, reflux, 12 h, nitrogen atmosphere (ii) Thiourea, dry toluene, reflux (iii) Chlorosulfonic acid, dry dichloromethane, 6 h	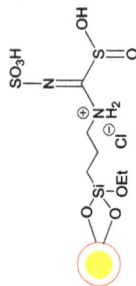	FT-IR spectroscopy: Bands at 2930 cm^{-1}: Propyl group stretching vibration of the C–H bonds at 3369 and 3212 cm^{-1}: Stretching of the N–H group on the thiourea dioxide at 1698 and 1636 cm^{-1}: Stretching vibrations of the C–N group on the imine moiety at 2700–3600 cm^{-1}: Vibrations related to O–H bonds in the SO$_3$H and SO$_2$H functional groups	[58]
Ionic liquid	Anion-DMIL, SMNP, toluene, reflux, 24 h (DMIL= 1,4-di(1H-imidazol-1-yl)butane)		FT-IR spectroscopy: 1475, 1538, and 1660 cm^{-1} correspond to C–H bending vibrations, C=C and C=N stretching vibrations of the imidazolium ring, peaks at 1542 and 1450 cm^{-1}, 1718 and 875 cm^{-1}, 827 cm^{-1}, 1025 cm^{-1}, 690 and 822 cm^{-1}, 1472, 1365 cm^{-1} correspond to vibrational modes of ZnCl$_3$, HSO$_4$, PF$_6$, CF$_3$SO$_3$, p-CH$_3$C$_6$H$_4$SO$_3$, and TiCl$_5$, respectively	[79]

(Continued)

Table 4.1. (*Continued*)

Modifying agent	Conditions	Structure of nanocomposite	Characterization	Ref.
Ionic liquid (1-butyl-3-(3-trimethoxypropyl)-1H-imidazol-3-ium chloride)	SMNP, ethanol (95%), 1-butyl-3-(3-trimethoxypropyl)-1H-imidazol-3-ium chloride, concentrated ammonia (28%), 36 h, N_2 atmosphere, room temperature		FT-IR spectroscopy: Signals at 1714 and 1544 cm^{-1} correspond to imidazole bonds	[80]
Ionic liquid [(MeO)$_3$SiPr MIM]PW	(i) SMNP, [(MeO)$_3$SiPr·MIM]Cl, dry toluene, 90°C, 24 h (ii) Dry CH$_2$Cl$_2$, ice bath, phospho-tungstic acid (H$_3$PW$_{12}$O$_{40}$), 0–5°C (iii) reflux, 48 h		FT-IR spectroscopy: Broad band around 1600 cm^{-1} (imidazole ring) 800–1000 cm^{-1} (Keggin-type heteropolyanions)	[81]
Ionic liquid 1-(tri-ethoxy silyl-propyl)-3-methyl-imidazolium hydrogen sulfate	(i) SiO$_2$ NPs dry toluene (ii) IL-HSO$_4$, nitrogen atmosphere, 90°C, 16 h		NMR Spectroscopy	[82]

N-heterocyclic carbene	Si@Fe$_3$O$_4$, [(NHC)Pd(allyl)Cl], 4 Å MS, toluene, reflux		TEM analysis	[83]
Carbene–palladium complex	(i) Imidazole, THF, NaH, inert atmosphere, 1 h, rt (ii) Tetrabutylammonium hydrogen sulfate, decyl bromide, THF, 24 h (iii) Ludox colloidal silica (Ludox-SM, 30 wt% SiO$_2$, pH 10), deionized water (iv) 1-decyl-3-(triethoxysilyl propyl)imidazolium chloride, MeOH (40 wt.%)		TEM analysis	[84]

(Continued)

Table 4.1. (*Continued*)

Modifying agent	Conditions	Structure of nanocomposite	Characterization	Ref.
Calix[4] resorcinarene	(i) SMNP, ethanol: Deionized water (1:1), reflux, 3-chloro propyltriethoxysilane, 24 h (ii) Calix[4]resorcinarene, toluene, potassium carbonate, tetrabutylammonium bromide, reflux, nitrogen atmosphere, 24 h		FT-IR spectroscopy: Peak in the range of 3500–3200 cm^{-1} corresponds to the hydroxyl stretching of internal phenol hydroxyl groups. The propyl groups attached to the resorcinarene framework in MNPs-Si-resorcinarene are identified by methylene stretching bands at 2959 and 2870 cm^{-1}	[85]
Salicylaldimine cobalt complex	(i) SNP, 3-APTES, dry toluene, reflux, 24 h, N$_2$ atmosphere (ii) Salicylaldehyde, methanol, reflux, 3 h		FT-IR spectroscopy: Band at 1637 cm^{-1} corresponds to C–N stretching frequency of the imine group (before complexation), new band around 1625 cm^{-1} after complexation indicates the C–N stretching frequency of the complexation	[86]

| Octadecyl | (i) SMNP, toluene, 80°C
 (ii) Trimethylamine, dimethyloctade-cylchlorosilane, reflux, 24 h | | FT-IR spectroscopy: Absorption peaks at 2920 and 2850 cm^{-1} represent the asymmetric and the symmetric vibrations of the $-CH_2$ group in the $-(CH_2)_{17}CH_3$ chain | [87] |
| Trisaccharide | SMNP, mixture of H$_2$O and MeOH, ammonia solution, dipropargyl derivative, ultrasonication, 3 h | | FT-IR spectroscopy: Peaks at 1714 and 1629 cm^{-1} correspond to the trisaccharide antigen structure, peak at 2100 cm^{-1} corresponds to asymmetric stretching of azido group | [88] |

Note: *Adapted from Ref. [68].

the method used to synthesize Mn complex was through a greener route, in addition to the preparation of catalyst and its application for oxidizing benzyl halides and alcohols. Employment of H_2O_2 as the sole and green oxidant, the solvent-free (or use of non-toxic ethanol as solvent) process, simple magnetic recovery and excellent yields with high turnover number are some other advantages that make this protocol a benign alternative for preparing carbonyl compounds [59].

Another green method was developed for the enzymatic degradation of ferulic acid, a model pollutant, through immobilization of soybean peroxidase onto a magnetic nanosupport by covalent attachment. The magnetite particles were characterized before and after chemical modification by XRD, SEM and TEM analysis. It was observed that the resulting nanobiocatalyst (enzyme load 5.25 U) was able to remove 99.67% of ferulic acid while free enzyme could remove only 57.67% under the same reaction conditions. Besides, the immobilized peroxidase could easily be recovered with magnetic forces and reused. High performance of the prepared MNPs suggests that the fabricated nanocatalyst can be used for pollutant removal from the environment [60].

Table 4.1 summarizes the common modifying strategies employed for the surface functionalization of silica-coated NPs. Figure 4.3 epitomizes various surface-modifying agents for core–shell SNPs.

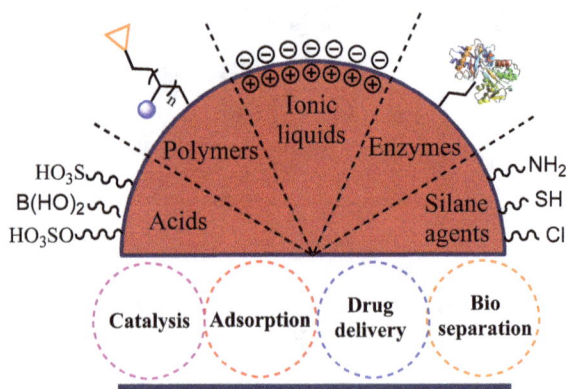

Figure 4.3.   Surface functionalization on silica-coated NPs and their applications.

4.4 Conclusion

In this chapter, we have summarized different strategies for surface functionalization of silica-coated NPs. Furthermore, various modifying agents including polymers, ionic liquids, biomolecules and others have been discussed along with their potential applications. It is evident from the examples discussed that functionalization techniques still present significant conceptual challenges, where most methods employed are multistep, tedious and work successfully only under specific conditions. Despite all the accomplishments, many new routes and applications are emerging. However, one of the most significant challenge is the lack of explicit modifying agents that are energy efficient, can be easily synthesized and are suitable for diverse applications.

References

[1] (a) C. Sanchez, P. Belleville, M. Popall, L. Nicole, *Chemical Society Reviews* **2011**, *40*, 696–753; (b) F. Zaera, *Chemical Society Reviews* **2013**, *42*, 2746–2762; (c) W. Wu, C. Jiang, V. A. Roy, *Nanoscale* **2015**, *7*, 38–58.

[2] V. Polshettiwar, D. Cha, X. Zhang, J. M. Basset, *Angewandte Chemie* **2010**, *122*, 9846–9850.

[3] R. K. Sharma, M. Yadav, M. B. Gawande, *Ferrites and Ferrates: Chemistry and Applications in Sustainable Energy and Environmental Remediation*, ACS Publications, **2016**, pp. 1–38.

[4] (a) R. K. Sharma, S. Dutta, S. Sharma, R. Zboril, R. S. Varma, M. B. Gawande, *Green Chemistry* **2016**, *18*, 3184–3209; (b) M. B. Gawande, Y. Monga, R. Zboril, R. Sharma, *Coordination Chemistry Reviews* **2015**, *288*, 118–143.

[5] (a) S. Laurent, D. Forge, M. Port, A. Roch, C. Robic, L. Vander Elst, R. N. Muller, *Chemical Reviews* **2008**, *108*, 2064–2110; (b) S. Shylesh, V. Schünemann, W. R. Thiel, *Angewandte Chemie International Edition* **2010**, *49*, 3428–3459; (c) L. M. Rossi, M. A. S. Garcia, L. L. R. Vono, *Journal of the Brazilian Chemical Society* **2012**, *23*, 1959–1971; (d) M. B. Gawande, P. S. Branco, R. S. Varma, *Chemical Society Reviews* **2013**, *42*, 3371–3393.

[6] (a) S. Giri, B. G. Trewyn, M. P. Stellmaker, V. S. Y. Lin, *Angewandte Chemie International Edition* **2005**, *44*, 5038–5044; (b) C.-W. Lu, Y. Hung, J.-K. Hsiao, M. Yao, T.-H. Chung, Y.-S. Lin, S.-H. Wu, S.-C. Hsu, H.-M. Liu, C.-Y. Mou, *Nano Letters* **2007**, *7*, 149–154; (c) T. Kim, E. Momin, J. Choi, K. Yuan, H. Zaidi, J. Kim, M. Park, N. Lee, M. T. McMahon, A. Quinones-Hinojosa, *Journal of the American Chemical Society* **2011**, *133*, 2955–2961; (d) H.-H. Yang, S.-Q. Zhang, X.-L. Chen,

Z.-X. Zhuang, J.-G. Xu, X.-R. Wang, *Analytical Chemistry* **2004**, *76*, 1316–1321.

[7] (a) N. Ž. Knežević, E. Ruiz-Hernández, W. E. Hennink, M. Vallet-Regí, *RSC Advances* **2013**, *3*, 9584–9593; (b) A. H. Lu, E. E. L. Salabas, F. Schüth, *Angewandte Chemie International Edition* **2007**, *46*, 1222–1244.

[8] C. Huang, B. Hu, *Spectrochimica Acta Part B: Atomic Spectroscopy* **2008**, *63*, 437–444.

[9] (a) A. Mobinikhaledi, H. Moghanian, M. Ghanbari, *Applied Organometallic Chemistry* **2018**, *32*, e4108–e4120; (b) X. Du, Q. He, L. Zhang, C. Liu, J. Zhu, B. Kuang, S. Zeng, B. Chen, D. Yin, Y. Zeng, *Bioconjugate Chemistry* **2017**, *28*, 2514–2517.

[10] F. Kashanian, G. Kokkinis, J. Bernardi, M. Zand, A. Shamloo, I. Giouroudi, *Sensors and Actuators A: Physical* **2017**, *270*, 223–230.

[11] O. Makrygenni, E. Secret, A. Michel, D. Brouri, V. Dupuis, A. Proust, J.-M. Siaugue, R. Villanneau, *Journal of Colloid and Interface Science* **2018**, *514*, 49–58.

[12] M. Aghababaie, M. Beheshti, A.-K. Bordbar, A. Razmjou, *RSC Advances* **2018**, *8*, 4561–4570.

[13] A. K. Tucker-Schwartz, R. A. Farrell, R. L. Garrell, *Journal of the American Chemical Society* **2011**, *133*, 11026–11029.

[14] R. He, X. You, J. Shao, F. Gao, B. Pan, D. Cui, *Nanotechnology* **2007**, *18*, 315601.

[15] V. O. Ikem, A. Menner, A. Bismarck, *Angewandte Chemie International Edition* **2008**, *47*, 8277–8279.

[16] Y.-H. Deng, C.-C. Wang, J.-H. Hu, W.-L. Yang, S.-K. Fu, *Colloids and Surfaces A: Physicochemical and Engineering Aspects* **2005**, *262*, 87–93.

[17] L. M. Rossi, L. Shi, F. H. Quina, Z. Rosenzweig, *Langmuir* **2005**, *21*, 4277–4280.

[18] G. Wang, Y. Sun, D. Li, H. W. Liang, R. Dong, X. Feng, K. Müllen, *Angewandte Chemie* **2015**, *127*, 15406–15411.

[19] Y. Lu, H. Fan, A. Stump, T. L. Ward, T. Rieker, C. J. Brinker, *Nature* **1999**, *398*, 223–226.

[20] R. K. Sharma, Y. Monga, A. Puri, G. Gaba, *Green Chemistry* **2013**, *15*, 2800–2809.

[21] (a) W. Posthumus, P. Magusin, J. Brokken-Zijp, A. Tinnemans, R. Van der Linde, *Journal of Colloid and Interface Science* **2004**, *269*, 109–116; (b) F. Torney, B. G. Trewyn, V. S.-Y. Lin, K. Wang, *Nature Nanotechnology* **2007**, *2*, 295–300.

[22] M. Dargahi-Zaboli, E. Sahraei, B. Pourabbas, B. A. Korgel, *Colloid and Polymer Science* **2017**, *295*, 925–932.

[23] X. Wang, Y. Zhang, W. Luo, A. A. Elzatahry, X. Cheng, A. Alghamdi, A. M. Abdullah, Y. Deng, D. Zhao, *Chemistry of Materials* **2016**, *28*, 2356–2362.

[24] I. J. Bruce, T. Sen, *Langmuir* **2005**, *21*, 7029–7035.

[25] W.-H. Chen, G.-F. Luo, Q. Lei, F.-Y. Cao, J.-X. Fan, W.-X. Qiu, H.-Z. Jia, S. Hong, F. Fang, X. Zeng, *Biomaterials* **2016**, *76*, 87–101.

[26] K. R. Hurley, H. L. Ring, M. Etheridge, J. Zhang, Z. Gao, Q. Shao, N. D. Klein, V. M. Szlag, C. Chung, T. M. Reineke, *Molecular Pharmaceutics* **2016**, *13*, 2172–2183.

[27] Q. Zhang, X. Wang, P. Z. Li, K. T. Nguyen, X. J. Wang, Z. Luo, H. Zhang, N. S. Tan, Y. Zhao, *Advanced Functional Materials* **2014**, *24*, 2450–2461.

[28] A. Popat, S. B. Hartono, F. Stahr, J. Liu, S. Z. Qiao, G. Q. M. Lu, *Nanoscale* **2011**, *3*, 2801–2818.

[29] F. Šulek, M. Drofenik, M. Habulin, Ž. Knez, *Journal of Magnetism and Magnetic Materials* **2010**, *322*, 179–185.

[30] P. Esmaeilnejad-Ahranjani, M. Kazemeini, G. Singh, A. Arpanaei, *Langmuir* **2016**, *32*, 3242–3252.

[31] V. Singh, K. Rakshit, S. Rathee, S. Angmo, S. Kaushal, P. Garg, J. H. Chung, R. Sandhir, R. S. Sangwan, N. Singhal, *Bioresource Technology* **2016**, *214*, 528–533.

[32] H. Rahma, R. Nickel, E. Skoropata, Y. Wroczynskyj, C. Rutley, P. K. Manna, C. H. Hsiao, H. Ouyang, J. van Lierop, S. Liu, *RSC Advances* **2016**, *6*, 65837–65846.

[33] H. Rahma, S. Asghari, S. Logsetty, X. Gu, S. Liu, *ACS Applied Materials & Interfaces* **2015**, *7*, 11536–11546.

[34] R. K. Singh, T. H. Kim, K. D. Patel, J. C. Knowles, H. W. Kim, *Journal of Biomedical Materials Research Part A* **2012**, *100*, 1734–1742.

[35] Y. Bai, Y. Cui, G. C. Paoli, C. Shi, D. Wang, M. Zhou, L. Zhang, X. Shi, *Colloids and Surfaces B: Biointerfaces* **2016**, *145*, 257–266.

[36] A. Otto, I. Mrozek, H. Grabhorn, W. Akemann, *Journal of Physics: Condensed Matter* **1992**, *4*, 1143–1212.

[37] F. Fu, Q. Wang, *Journal of Environmental Management* **2011**, *92*, 407–418.

[38] G.-Z. Fang, J. Tan, X.-P. Yan, *Analytical Chemistry* **2005**, *77*, 1734–1739.

[39] R. Rostamian, M. Najafi, A. A. Rafati, *Chemical Engineering Journal* **2011**, *171*, 1004–1011.

[40] A. Heidari, H. Younesi, Z. Mehraban, *Chemical Engineering Journal* **2009**, *153*, 70–79.

[41] R. Sharma, A. Puri, Y. Monga, A. Adholeya, *Separation and Purification Technology* **2014**, *127*, 121–130.

[42] R. Sharma, A. Puri, Y. Monga, A. Adholeya, *Journal of Materials Chemistry A* **2014**, *2*, 12888–12898.

[43] M. H. P. Wondracek, A. O. Jorgetto, A. C. P. Silva, J. do Rocio Ivassechen, J. F. Schneider, M. J. Saeki, V. A. Pedrosa, W. K. Yoshito, F. Colauto, W. A. Ortiz, *Applied Surface Science* **2016**, *367*, 533–541.

[44] M. O. Ojemaye, O. O. Okoh, A. I. Okoh, *Separation and Purification Technology* **2017**, *183*, 204–215.

[45] R. Roto, Y. Yusran, A. Kuncaka, *Applied Surface Science* **2016**, *377*, 30–36.

[46] M. Zarei, A. Shahpiri, P. Esmaeilnejad-Ahranjani, A. Arpanaei, *RSC Advances* **2016**, *6*, 46785–46793.

[47] T. Lü, S. Zhang, D. Qi, D. Zhang, H. Zhao, *Journal of Alloys and Compounds* **2016**, *688*, 513–520.

[48] D. C. Culita, C. M. Simonescu, R.-E. Patescu, M. Dragne, N. Stanica, O. Oprea, *Journal of Solid State Chemistry* **2016**, *238*, 311–320.

[49] F. Latifeh, Y. Yamini, S. Seidi, *Environmental Science and Pollution Research* **2016**, *23*, 4411–4421.

[50] D. Hughes, A. Afsar, L. M. Harwood, T. Jiang, D. M. Laventine, L. J. Shaw, M. E. Hodson, *Chemosphere* **2017**, *183*, 519–527.

[51] S. B. Kalidindi, B. R. Jagirdar, *ChemSusChem* **2012**, *5*, 65–75.

[52] R. Sharma, S. Sharma, G. Gaba, S. Dutta, *Journal of Materials Science* **2016**, *51*, 2121–2133.

[53] R. Sharma, S. Sharma, *Dalton Transactions* **2014**, *43*, 1292–1304.

[54] S. K. Dangolani, F. Panahi, M. Nourisefat, A. Khalafi-Nezhad, *RSC Advances* **2016**, *6*, 92316–92324.

[55] M. Gholinejad, A. Neshat, F. Zareh, C. Nájera, M. Razeghi, A. Khoshnood, *Applied Catalysis A: General* **2016**, *525*, 31–40.

[56] A. Bayat, M. Shakourian-Fard, N. Talebloo, M. M. Hashemi, *Applied Organometallic Chemistry* **2018**, *32*, e4061–e4070.

[57] S. Swami, A. Agarwala, R. Shrivastava, *New Journal of Chemistry* **2016**, *40*, 9788–9794.

[58] M. A. Zolfigol, R. Ayazi-Nasrabadi, *RSC Advances* **2016**, *6*, 69595–69604.

[59] R. K. Sharma, M. Yadav, Y. Monga, R. Gaur, A. Adholeya, R. Zboril, R. S. Varma, M. B. Gawande, *ACS Sustainable Chemistry & Engineering* **2016**, *4*, 1123–1130.

[60] M. C. Silva, J. A. Torres, F. G. Nogueira, T. S. Tavares, A. D. Corrêa, L. C. Oliveira, T. C. Ramalho, *RSC Advances* **2016**, *6*, 83856–83863.

[61] A. Rostami, A. Ghorbani-Choghamarani, B. Tahmasbi, F. Sharifi, Y. Navasi, D. Moradi, *Journal of Saudi Chemical Society* **2017**, *21*, 399–407.

[62] M. Zolfigol, A. Khazaei, M. Mokhlesi, F. Derakhshan-Panah, *Journal of Molecular Catalysis A: Chemical* **2013**, *370*, 111–116.

[63] D. Azarifar, O. Badalkhani, Y. Abbasi, *Journal of Sulfur Chemistry* **2016**, *37*, 656–673.

[64] A. Yari, S. Rashnoo, *Journal of Chromatography B* **2017**, *1067*, 38–44.

[65] Y.-Y. Lee, K. C.-W. Wu, *Physical Chemistry Chemical Physics* **2012**, *14*, 13914–13917.

[66] L. Torkian, P. Salehi, M. Dabiri, S. Kharrazi, *Synthetic Communications* **2011**, *41*, 2115–2122.

[67] F. Porta, G. E. Lamers, J. Morrhayim, A. Chatzopoulou, M. Schaaf, H. den Dulk, C. Backendorf, J. I. Zink, A. Kros, *Advanced Healthcare Materials* **2013**, *2*, 281–286.

[68] E. R. Riva, I. Pastoriza-Santos, A. Lak, T. Pellegrino, J. Pérez-Juste, V. Mattoli, *Journal of Colloid and Interface Science* **2017**, *502*, 201–209.

[69] J. Noh, S. Hong, C.-M. Yoon, S. Lee, J. Jang, *Chemical Communications* **2017**, *53*, 6645–6648.

[70] D. F. Enache, E. Vasile, C. M. Simonescu, A. Răzvan, A. Nicolescu, A.-C. Nechifor, O. Oprea, R.-E. Pătescu, C. Onose, F. Dumitru, *Journal of Solid State Chemistry* **2017**, *253*, 318–328.

[71] M. Behbahani, A. Veisi, F. Omidi, A. Noghrehabadi, A. Esrafili, M. H. Ebrahimi, *Applied Organometallic Chemistry* **2018**, *32*, e4133–e4134.

[72] S. Abdulhussain, H. Breitzke, T. Ratajczyk, A. Grünberg, M. Srour, D. Arnaut, H. Weidler, U. Kunz, H. J. Kleebe, U. Bommerich, *Chemistry — A European Journal* **2014**, *20*, 1159–1166.

[73] M. Masteri-Farahani, M. Modarres, *Monatshefte für Chemie — Chemical Monthly* **2017**, *148*, 1403–1410.

[74] F. Farzaneh, Y. Sadeghi, M. Maghami, Z. Asgharpour, *Journal of Cluster Science* **2016**, *27*, 1701–1718.

[75] H. Zhou, H. Wang, H. Niu, A. Gestos, T. Lin, *Advanced Functional Materials* **2013**, *23*, 1664–1670.

[76] M. A. Zolfigol, M. Kiafar, M. Yarie, A. A. Taherpour, M. Saeidi-Rad, *RSC Advances* **2016**, *6*, 50100–50111.

[77] F. A. Casado-Carmona, M. del Carmen Alcudia-León, R. Lucena, S. Cárdenas, M. Valcárcel, *Microchemical Journal* **2016**, *128*, 347–353.

[78] H. Abdolmohammad-Zadeh, S. Hassanlouei, M. Zamani-Kalajahi, *RSC Advances* **2017**, *7*, 23293–23300.

[79] Y. L. Hu, R. Xing, *Catalysis Letters* **2017**, *147*, 1453–1463.

[80] G. Rahimzadeh, S. Bahadorikhalili, E. Kianmehr, M. Mahdavi, *Molecular Diversity* **2017**, *21*, 597–609.

[81] M. H. Ghasemi, E. Kowsari, *Research on Chemical Intermediates* **2017**, *43*, 1957–1968.

[82] K. B. Sidhpuria, A. L. Daniel-da-Silva, T. Trindade, J. A. Coutinho, *Green Chemistry* **2011**, *13*, 340–349.

[83] J.-M. Collinson, J. D. Wilton-Ely, S. Díez-González, *Catalysis Communications* **2016**, *87*, 78–81.

[84] S. Tandukar, A. Sen, *Journal of Molecular Catalysis A: Chemical* **2007**, *268*, 112–119.

[85] A. Mouradzadegun, S. Boroon, P. K. Fard, *Monatshefte für Chemie — Chemical Monthly* **2017**, *148*, 367–374.

[86] D. Tang, L. Zhang, Y. Zhang, Z.-A. Qiao, Y. Liu, Q. Huo, *Journal of Colloid and Interface Science* **2012**, *369*, 338–343.

[87] W. Kaewsuwan, P. Kanatharana, O. Bunkoed, *Journal of Analytical Chemistry* **2017**, *72*, 957–965.

[88] S. Yan, C. Zhao, Q. Ren, X. Xie, F. Yang, Y. Du, *Tetrahedron* **2017**, *73*, 2949–2955.

[89] X. Kong, Q. Yu, X. Zhang, X. Du, H. Gong, H. Jiang, *Journal of Materials Chemistry*, **2012**, *22(16)*, 7767–7774.

Chapter 5

Characterization of Metal-Immobilized Silica Nanoparticles and Silica-Coated Magnetic Nanoparticles

Rashmi Gaur, Shivani Sharma, Yukti Monga
and Rakesh Kumar Sharma*
*rksharmagreenchem@hotmail.com

5.1 Introduction

Since the beginning of catalytic research, improvement and fabrication of heterogeneous catalyst have become more and more significant [1]. However in the early 1950s, only a few techniques were available for the characterization of synthesized catalysts which provided information about the chemical composition, structure (X-ray diffraction (XRD)) and texture analysis of the specific surface area and pore size distribution (Brunauer–Emmette–Teller (BET)) [2]. All these data seemed insufficient to determine the specific conditions required for the catalyst preparation and consequently their mode of action. Moreover, during that period, various discussions often turned acrimonious due to the insufficient knowledge of the catalytic materials. It was found that information about the synthesized material is indispensable for determining the type of catalysts, the scope of improvement of known catalysts and the advancement of new ones [3]. In later years, new techniques were developed which

---

*Corresponding author.

have been rapidly adopted for better understanding of fabricated catalysts [4].

Material characterization is the practice of identifying and measuring physical, mechanical, chemical and microstructural properties of materials [5]. After the complete characterization of synthesized materials, many important issues such as source of failure, identification of contaminants and adsorbed species and analysis of each step in synthetic methodology can be resolved. However, each characterization technique has its own specific limitations. It is indispensable to employ more than one technique for inclusive evaluation of a fabricated catalyst [6]. Several physicochemical characterization techniques, such as solid-state ^{13}C CPMAS and ^{29}Si CPMAS nuclear magnetic resonance (NMR) spectroscopy, Fourier transform infrared spectroscopy (FTIR), XRD analysis, transmission electron microscopy (TEM), scanning electron microscopy (SEM), atomic force microscopy (AFM), thermogravimetric analysis (TGA), X-ray photoelectron spectroscopy (XPS), energy-dispersive X-ray fluorescence spectroscopy (EDXRF), vibrating sample magnetometry (VSM), atomic absorption spectroscopy (AAS) and BET analysis, are employed in order to authenticate the synthesis of silica nanoparticles (SNPs) and silica-coated magnetic nanoparticles (SMNPs)-based materials.

In this chapter, detailed characterizations of SNPs and SMNPs shall be discussed (Table 5.1).

5.2 Transmission Electron Microscopy

TEM is a technique whereby a beam of electrons is transmitted through an ultrathin specimen in order to get information about the sample's shape, size, morphology, core–shell formation by contrast difference and coating thickness. In order to see resolution even at molecular scale, high-resolution TEM (HRTEM) is used and measures the average lattice interfringe distance, which corresponds to one of the planes of inverse spinel structured of magnetic nanoparticles (MNPs) core. The selected area electron diffraction (SAED) image of MNPs confirms the crystallinity through the presence of

Table 5.1. Major analytical techniques employed for the characterization of functionalized SNPs and SMNPs.

Analytical parameter	Technique employed	Information extracted
Thermal stability	TGA	Gives information about the thermal stability of synthesized materials.
Magnetic property	VSM	Provides information about the magnetic properties of the material.
Metal loading	AAS/ICP-OES	Determines the metal content on the surface of nanosupport.
	EDS	Provides qualitative and quantitative information about the metal present on the nanosupport.
Structure	FTIR	Confirms the stepwise modification of the synthesized nanocomposites through identification of appropriate functional groups.
	XRD	Provides information about the crystallographic structure, chemical composition of the nanocomposites.
	NMR	Confirms the covalent grafting of the organic functionalities onto silica NPs surface. Silica-coated magnetic nanoparticles are inactive for NMR spectroscopy.
Surface properties	XPS	Provides qualitative information, chemical status and oxidation state of element present in the nanocatalyst.
	STEM	Offers spatial distribution of elements.
	TEM/SEM	Provides information about size and shape of the synthesized nanoparticles.
	BET	Measures the surface area of the catalytic materials.
	AFM	Provides the diameter and height of the particles.

bright diffraction rings. The crucial step of silica coating may also be confirmed with TEM image, through the contrast difference between the dark magnetite nanocores surrounded by a light grey silica shell. However, this technique is incompetent to generate three-dimensional images of nanoparticles (NPs). The sample preparation includes dispersion of the NPs with the help of ultrasonication into absolute ethanol and casting a drop of this suspension on the carbon-coated copper TEM grids.

Sharma and co-workers [7] reported the preparation of manganese-based silica-coated magnetic nanocatalyst and characterized it by TEM. From HRTEM images of MNPs, interplanar distance was observed to be 0.25 nm which corresponded to (311) plane of MNPs (Figure 5.1(a)). The obtained MNPs were spherical in shape and slight agglomeration was observed (Figure 5.1(b)). The SAED pattern of MNPs indicated the highly crystalline nature of Fe_3O_4 due to the presence of white spots and bright diffraction rings (Figure 5.1(c)). Figure 5.1(d) displayed silica coating over the MNPs core which confirmed the synthesis of SMNPs. The size of final catalyst and recovered nanocatalyst was found to be 30–40 nm, as shown in Figures 5.1(e) and 5.1(f) [7].

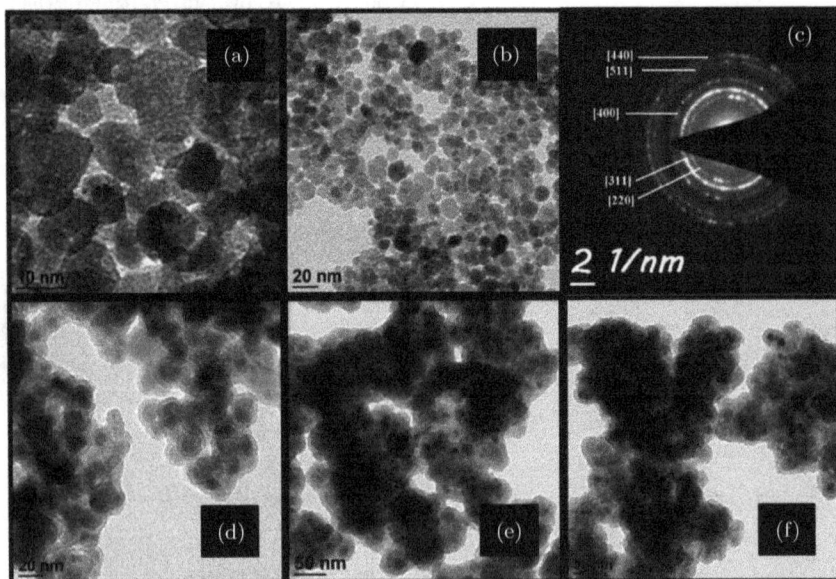

Figure 5.1.   (a) HRTEM image of MNPs, (b) TEM image of MNPs, (c) SAED pattern of MNPs, (d) TEM image of SMNPs, (e) final catalyst Mn−Ac@ASMNP and (f) recovered nanocatalyst (adapted from Ref. [7]).

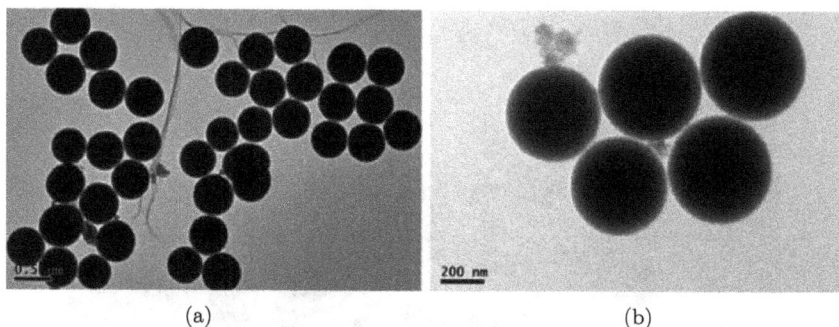

Figure 5.2. TEM micrographs of (a) SNSs and (b) SNS-supported palladium catalyst (adapted from Ref. [8]).

TEM micrographs of silica nanospheres (SNSs)-based materials have been provided in Figure 5.2. These micrographs reveal that the particles are monodisperse and spherical in shape. Also, they do not undergo agglomeration which accounts for their better catalytic activity [8].

5.3 Scanning Electron Microscopy

SEM is the most familiar microscopic technique utilized for determining the shape, size and roughness of nanoparticles. However, this technique is ineffective to differentiate between the core and shell materials because it can only create the surface image of nanoparticles. Recently, high-resolution image can also be generated by using field emission scanning electron microscopy (FESEM). The procedure for conducting SEM includes mounting the dry powder of the nanomaterial on a sample holder followed by gold coating using a sputter coater and then scanning the surface of the nanomaterial with a focused beam of electrons.

A SEM image of the Fe_3O_4@polypyrrole composite was reported recently which showed that the MNPs possess spherical morphology (Figure 5.3) [9]. Similarly, Sharma and co-workers have reported FESEM images of MNPs and Ag–AcPy@ASMNPs as shown in Figure 5.4 are spherical in shape, but FESEM monograph of final catalyst Ag–AcPy@ASMNPs reveals that the surface of MNPs

Figure 5.3. SEM image of Fe_3O_4@polypyrrole composite (adapted from Ref. [9]).

was roughened due to silica coating and functionalization of MNP (Figures 5.4(a) and 5.4(b)) [10]. On the contrary, Figure 5.5 displays the SEM images of uniform and spherical mesoporous SNPs having diameters of approximately 600 and 150 nm, respectively [11].

5.4 X-ray Diffraction

XRD is a non-destructive analytical technique that exhibits information about nanocomposites, such as the crystallographic structure and identification of unknown materials. Every compound exhibits a fingerprint characteristic XRD pattern that can be matched against

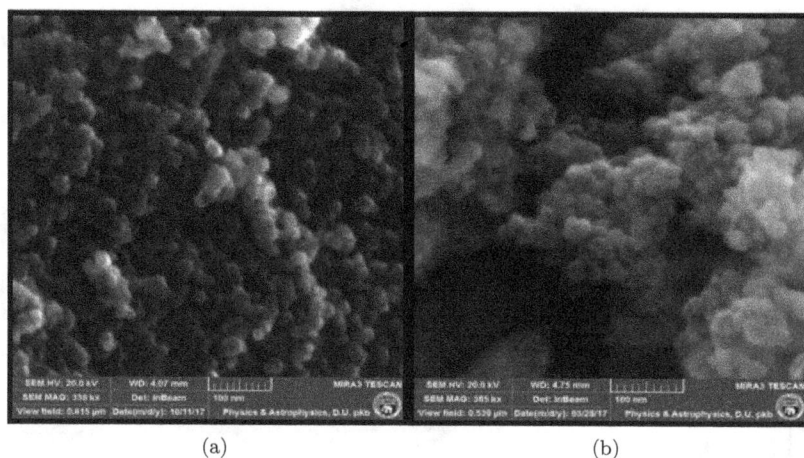

(a)   (b)

Figure 5.4. FESEM images of (a) MNPs and (b) Ag–AcPy@ASMNPs nanocatalyst (adapted from Ref. [10]).

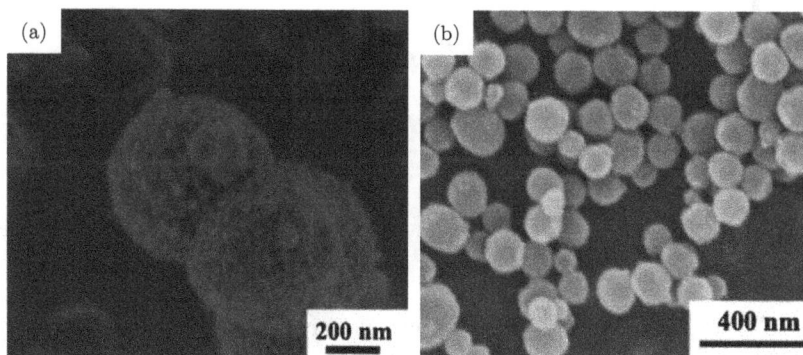

Figure 5.5. SEM images of (a) large pores and (b) small pores mesoporous SNPs (adapted from Ref. [11]).

the standard XRD data of Joint Committee on Powder Diffraction Standards (JCPDS). Recently, a group reported XRD pattern of MNPs, SMNPs and final catalyst AcTp@ASMNPs (Figure 5.6). MNPs exhibit the characteristic peaks of cubic inverse spinel structure and no extra peak was present, which denotes the high phase purity of the crystalline magnetite nanoparticles. The positions and the relative intensities in the diffractogram matched well with the

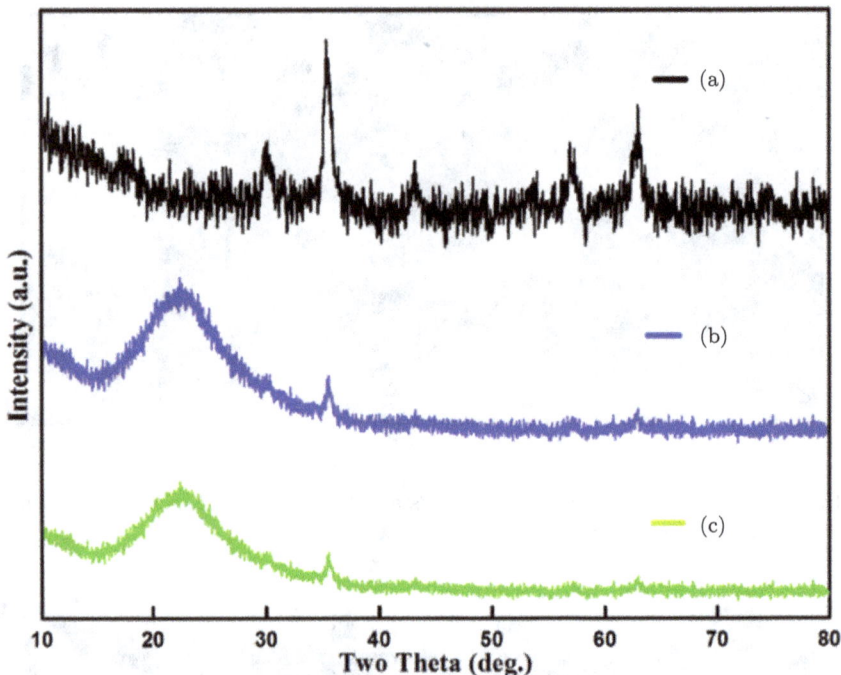

Figure 5.6.   XRD pattern of the (a) MNPs, (b) SMNPs and (c) AcTp@ASMNPs (adapted from Ref. [12]).

standard XRD data of JCPDS card number (19-0629) of crystalline magnetite nanoparticles (Figure 5.6(a)). The mean crystalline sizes of MNPs were calculated by Debye–Scherrer equation and found to be 12.9 nm. The XRD pattern for SMNPs showed the appearance of a weak broad band at $2\theta = 20$–$24°$ which is due to the amorphous silica shell formed around the magnetic core and confirms the silica encapsulation (Figure 5.6(b)). Figure 5.6(c) demonstrates the XRD pattern of ligand (AcTp) immobilized on amino-modified silica-coated magnetic nanoparticles (ASMNPs) and it was observed that diffraction peaks of magnetic core disappear completely due to sufficient or thick silica coating [12].

Similarly, Sharma *et al.* [13] presented the XRD patterns of SiO_2, SiO_2@APTES and SiO_2@APTES@DAFO-Fe, as shown in Figure 5.7. All silica-based nanomaterials exhibit a single broad diffraction peak

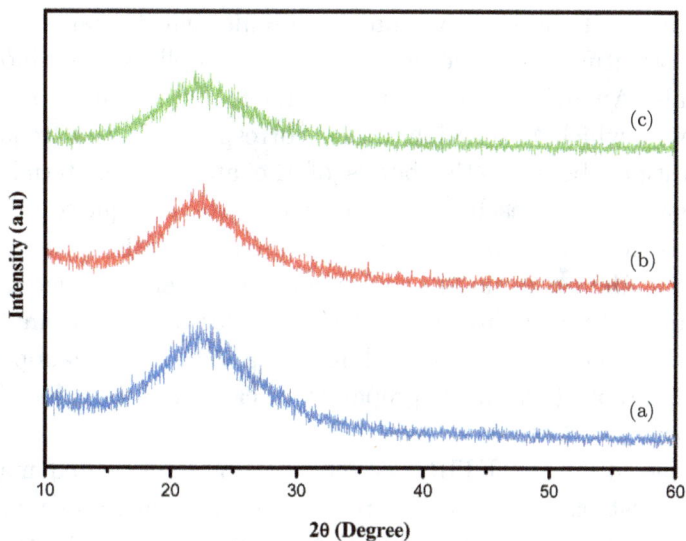

Figure 5.7. XRD diffraction patterns of (a) SiO_2, (b) SiO_2@APTES and (c) SiO_2@APTES@DAFO-Fe (adapted from Ref. [13]).

at $2\theta = 23°$ which is characteristic of the amorphous nature of the Stöber's SNSs. The results show that there is no remarkable change observed in the topological structure of SNSs even after the surface functionalization reactions. Hence, it may be concluded that the bare SNSs are stable enough to experience the chemical modification reactions starting from the surface functionalization to the binding of iron. However, upon comparison of the diffraction patterns of these samples, a decrease in the XRD peak intensities of SiO_2@APTES and SiO_2@APTES@DAFO-Fe is observed with the successive anchoring of APTES and iron complex onto SNSs. On the contrary, XRD analysis of nanomaterials does not show any novel diffraction peak. This gives a clear indication that these species existed on the surface of SNSs in the form of non-crystalline state [13].

5.5   Fourier Transform Infrared Spectroscopy

Infrared spectroscopy is an important technique in organic chemistry that is used to identify the presence of certain functional groups in the sample. It basically provides specific information about the vibrations and rotations within a molecule, which makes it useful for investigating the structure of organic as well as some inorganic materials. An infrared spectrum is said to be the fingerprint of a sample in which absorption peaks correspond to the frequencies of vibrations between the bonds of the atoms of material under investigation. Since each material is made up of a unique combination of atoms, no two compounds can exhibit exactly the same infrared spectrum. Hence, infrared spectroscopy has been considered as an excellent tool for qualitative analysis of different kinds of materials. Here, the functionalization and modification of the nanoparticles at each step of catalyst preparation can be monitored by IR spectroscopy.

In the literature, FTIR was reported for silica-coated magnetic nanocatalyst, as shown in Figure 5.8 [14]. The purity of magnetite phase can be determined by the presence of Fe–O stretching

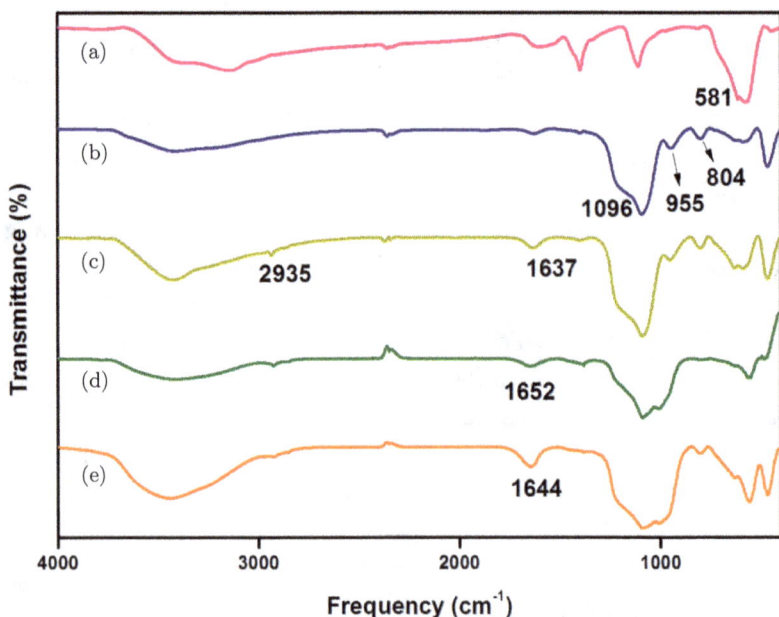

Figure 5.8. FTIR spectra of (a) MNPs, (b) SMNPs, (c) ASMNPs, (d) Sc@ASMNPs and (e) Pd–Sc@ASMNPs (adapted from Ref. [14]).

vibrations at 581 cm^{-1} (Figure 5.8(a)). The silica coating of magnetite nanoparticles (SMNPs) can be confirmed by the appearance of a broad band centered around 804, 955 and 1096 cm^{-1} which can be assigned to Si–O–Si symmetric, Si–O symmetric and Si–O–Si asymmetric stretching vibrations, respectively (Figure 5.8(b)). The bands at 2935 cm^{-1} and 1637 cm^{-1} correspond to the –CH$_2$ and –NH$_2$ groups of aminopropyl moiety of APTES that confirmed the synthesis of ASMNPs (Figure 5.8(c)). The spectrum of the ligand grafted over ASMNPs (Sc@ASMNPs) showed a strong band at 1652 cm^{-1} due to C=N stretching vibration (Figure 5.8(d)). On metalation, the vibration peak at 1652 cm^{-1} shifted to 1644 cm^{-1}, confirming the successful immobilization of palladium onto the surface of Sc@ASMNPs.

As an illustrative example, we have depicted the FTIR spectral assignments of nanomaterials, viz. SiO$_2$, SiO$_2$@APTES and SiO$_2$@APTES@Pd-FFR (Figure 5.9) [8]. The spectra show typical vibration bands of silica materials, such as the Si–O–Si asymmetric stretching band at 1095 cm^{-1} and Si–O symmetric stretching band at 805 cm^{-1}. Further, the aliphatic –CH$_2$ stretching vibration bands of the propyl chain of the silylating agent (APTES) appear at 2927 cm^{-1} and 2855 cm^{-1} in the IR spectrum of SiO$_2$@APTES, that are absent in the case of the SNSs, which clearly confirms that APTES has been anchored on the SNSs. It should be noted that the band corresponding to the Si–OH group which appears at 960 cm^{-1} is found to be a little weaker in the spectrum of SiO$_2$@APTES compared to that in the spectrum of the SNSs. The reason behind this is the reaction between the surface Si–OH groups of SiO$_2$ and ethoxy groups of APTES during the modification reaction. It is also worth noting that there are no significant changes observed in the SiO$_2$ structure-sensitive vibrations after its modification with APTES. Hence, it may be inferred that the basic silica framework remains intact. The spectrum of the SiO$_2$@APTES@Pd–FFR nanocatalyst shows a large number of bands which merely appear as shoulders because of their superimposition by a broad polymer band at 1088 cm^{-1}. In addition, the expected vibration band of the azomethine group appears at 1629 cm^{-1}. Thus, the covalent

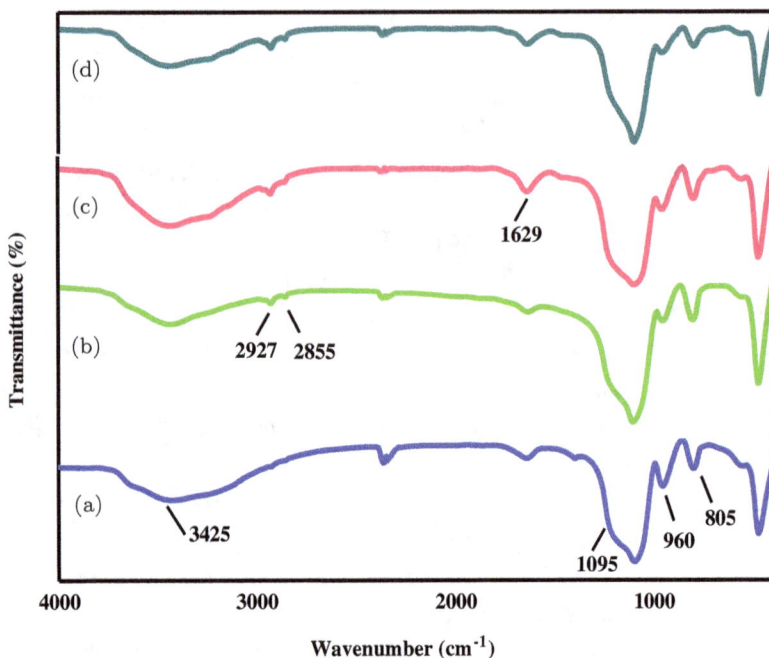

Figure 5.9. FTIR spectra of (a) SiO_2, (b) $SiO_2@APTES$, (c) $SiO_2@APTES@Pd$–FFR and (d) recovered $SiO_2@APTES@Pd$–FFR (adapted from Ref. [8]).

grafting of APTES on SiO_2 and the successive immobilization of furfuraldehyde (FFR) on $SiO_2@APTES$ are confirmed by FTIR spectroscopy.

5.6 Vibrating Sample Magnetometer

Magnetism is a remarkable property of silica-coated magnetic nanocatalyst. The superparamagnetic nature may be detected by VSM analysis through the curves of both magnetization and demagnetization that pass through the origin and display negligible and zero coercivity and remanence. In most of the cases with magnetite core, coercivity and remanence are negligible or zero. Encapsulation of the magnetite nanoparticles with silica induces lower magnetization value. Figure 5.10 indicates the VSM curves of MNPs,

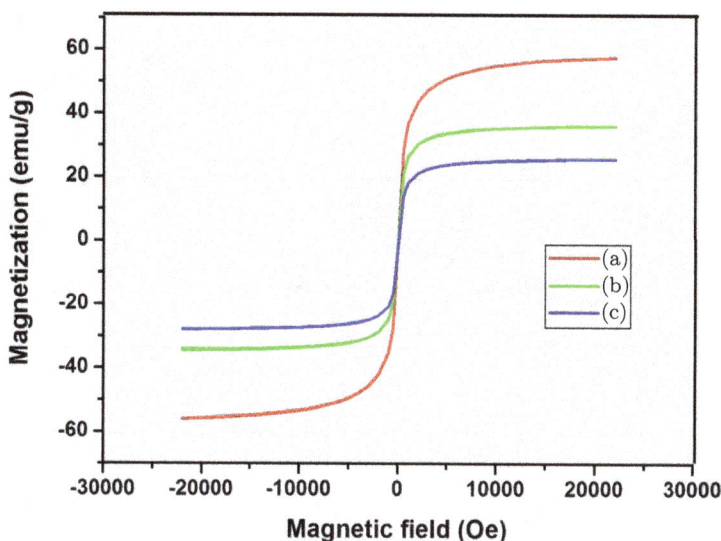

Figure 5.10. Magnetization curves obtained at room temperature for (a) MNPs, (b) SMNPs and (c) Ag–AcPy@ASMNPs (adapted from Ref. [10]).

SMNPs and Ag–AcPy@ASMNPs nanocatalyst. The curves reveal the decrease in the saturation magnetization (M_s) values from MNPs (58 emu/g) to SMNPs (36 emu/g) to the final Ag–AcPy@ASMNPS nanocatalyst (26 emu/g) (Figures 5.10(a)–5.10(c)) [10]. The variation between them is ascribed to the increased mass of diamagnetic silica over MNPs and immobilization of metal–ligand complex over functionalized SMNPs.

Another example of using VSM is to study the magnetization of Fe@Au nanoparticles as the magnetic field of susceptometer cycles between +60 and −60 kOe at 2, 10 and 300 K, respectively. It is clear from Figure 5.10, below the blocking temperature (42 K) Fe@Au nanoparticles are ferromagnetic in nature, displaying the remanence and coercivity of 4.12 emu/g and 728 Oe at 2 K (Figure 5.11(a)) and 2.92 emu/g and 322 Oe (Figure 5.11(b)) at 10 K, respectively. On the contrary, above the blocking temperature or at 300 K, no coercivity or remanence is observed, which confirms that MNPs are superparamagnetic at room temperature (300 K) (Figure 5.11(c)) [15].

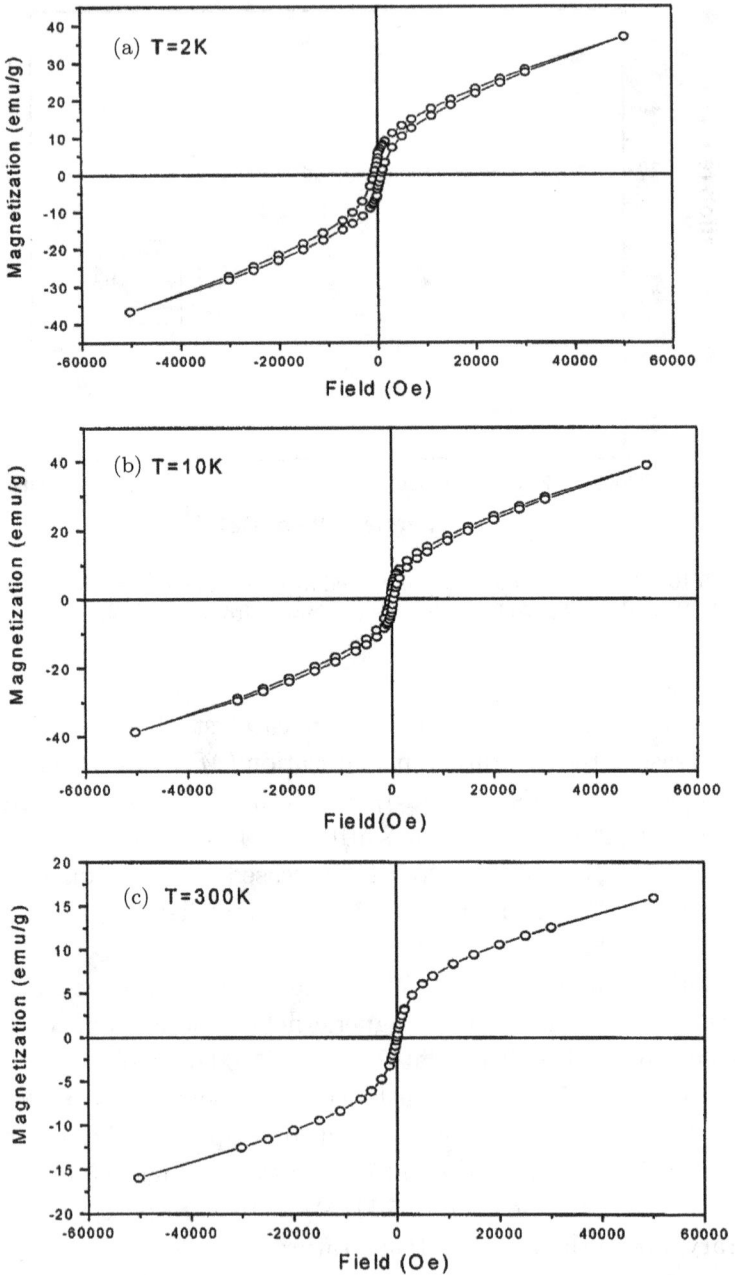

Figure 5.11. Magnetization curves of Fe@Au nanoparticles obtained at (a) 2 K, (b) 10 K and (c) 300 K (adapted from Ref. [15]).

5.7 Energy-Dispersive X-ray Spectroscopy

The compositional evidence for the presence of various elements at every step of catalyst preparation can be provided by the energy-dispersive X-ray spectroscopy (EDS) analysis. It can also give the pivotal evidence of the metal immobilization on the fabricated nanosupport. Elemental analysis can be carried out when SEM and TEM are coupled with EDX. Sharma and co-workers [12] prepared Cu(II)–AcTp@Am–Si–Fe$_3$O$_4$ magnetic nanocatalyst and studied it by using EDS analysis. Figure 5.12(a) displays the presence of silica and iron which indicates the encapsulation of magnetite core with silica, whereas copper immobilization over AcTp@Am–Si–Fe$_3$O$_4$ nanosupport is also verified by the EDS technique, as shown in Figure 5.12(b) [12].

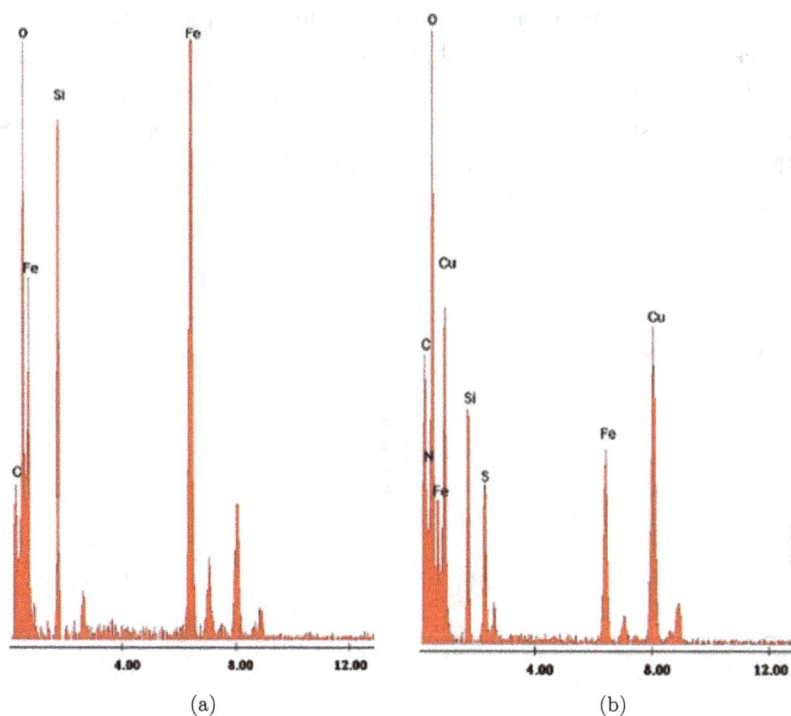

(a)   (b)

Figure 5.12. EDS pattern of (a) Si–Fe$_3$O$_4$ and (b) Cu–AcTp@Am–Si–Fe$_3$O$_4$ nanocatalyst (adapted from Ref. [12]).

5.8   X-ray Photoelectron Spectroscopy

XPS is another essential spectroscopic technique which is able to reveal the surface information, such as elemental composition, chemical status, empirical formula and oxidation state. An example of using XPS spectroscopy to examine final and reused nanocatalyst sample material is shown in Figure 5.13. Both the figures indicate the presence of $Cu2p_{1/2}$ and $Cu2p_{3/2}$ band at 953.6 and 933.4 eV, respectively, and the broad consequent satellite bands at 940.9, 943.4 and 962.1 eV further verify the presence of copper having (+2) oxidation state (Figures 5.13(a) and 5.13(b)) [16].

5.9   Thermal Gravimetric Analysis

TGA technique is used to describe the thermal stability of sample materials and offers a thermal profile of any sample material as the temperature varies. An example of using TGA technique to examine the MNPs and SMNPs is revealed in Figure 5.14 [17]. The TGA curves demonstrate a weight loss of 8% and 11% over 400°C in S1 (MNPs) and S2 (SMNPs), respectively, which is attributed to the removal of absorbed H_2O and other species.

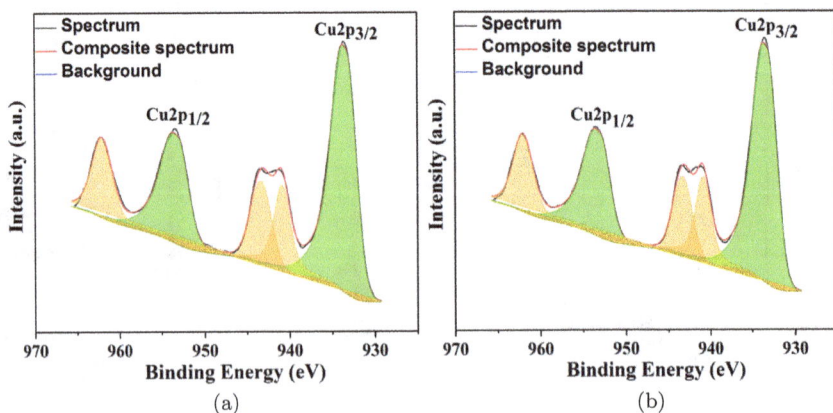

Figure 5.13. XPS spectra of Cu–ABF@ASMNPs catalyst: (a) fresh and (b) reused (adapted from Ref. [16]).

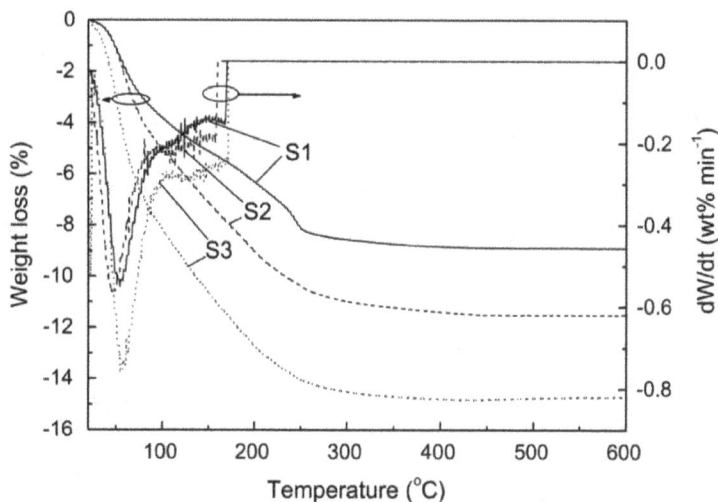

Figure 5.14. TGA curves of MNPs (S1) and SMNPs (S2) (adapted from Ref. [17]).

An efficient heterogeneous Pd catalytic system was fabricated based on the immobilization of Pd nanoparticles (PNPs) on a silica-bonded propylamine–cyanuric–cysteine (SiO$_2$–pA–Cyan–Cys) substrate and subsequently employed in the Suzuki and Sonogashira cross-coupling reactions by Ghiaci and group. Further, TGA of the catalyst was carried out by raising its temperature at a rate of 5°C min^{-1} in air up to 800°C to analyze its thermal decomposition performance. As is evident from Figure 5.15, two main stages of weight loss were observed. The first small weight loss below 250°C is attributed to the loss of free water and the second weight loss at about 400°C indicates decomposition of pA–Cyan–Cys–Pd [18].

5.10 BET Surface Area Analysis

The BET technique which is an extension of the Langmuir theory was developed by Stephen Brunauer, Paul Emmett and Edward Teller in the year 1938. It is the most common method utilized to measure the specific surface area of samples in units of area per mass of sample (m^2/g). This method has been widely used

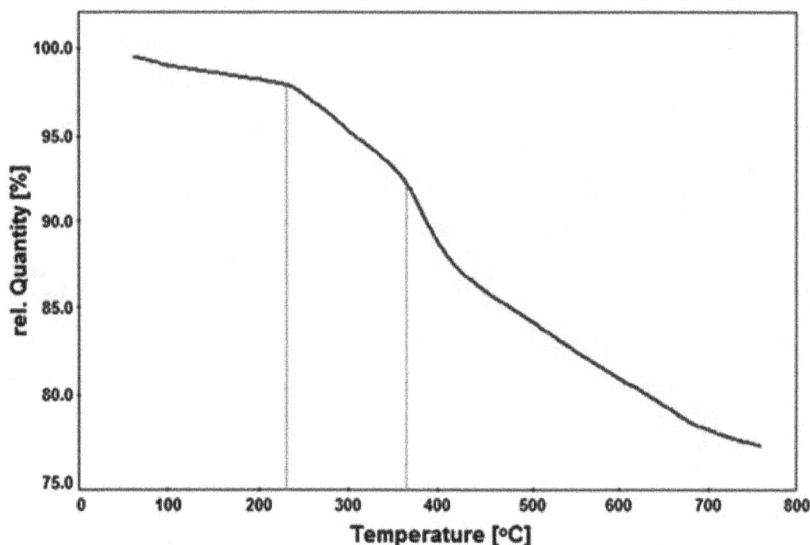

Figure 5.15.   TGA of pA–Cyan–Cys–Pd catalyst (adapted from Ref. [18]).

to determine the specific surface area of different materials for various applications, such as catalysis, particle and gas filtration, fuel cell technology, adsorption and many more. Gogoi and co-workers synthesized nanosilica-anchored Pd (II) Schiff base complex and utilized it as an efficient heterogeneous catalyst for activation of aryl halides in the Suzuki–Miyaura cross-coupling reaction in water [19]. The successive functionalization of the developed catalyst was affirmed using the BET surface area technique and the results have been presented in Table 5.2. As shown, there is a gradual decrease in surface area from nanosilica to anchored Pd(II) Schiff base complex which indicates the immobilization of the complex onto the nanosilica surface.

Similarly, Kalantary and co-workers explained in their report that SMNPs provides larger surface area than bare MNPs (Table 5.3) [20]. The enhancement of the surface area of SMNPs can be attributed to special textural structure and high pore size area. This property of SMNPs is beneficial from both catalytic and adsorption capacity viewpoints.

Table 5.2. BET surface area measurement of nanosilica-based materials.

S.No.	Material	S_{BET} (m^2g^{-1})
1.	Nanosilica	189.97
2.	APTES@nanoSiO$_2$	175.45
3.	Schiff base anchored to nanosilica	169.21
4.	Pd(II) Schiff base complex	160

Table 5.3. Surface area of MNPs and SMNPs.

Material	$S_{BET}(m^2/g)$	Langmuir specific surface area (m^2/g)	Pore volume (cc/g)	Average pore diameter (nm)
MNPs	98.7	118.3	2.58	3.1
SMNPs	272.5	296.4	4.7	3.7

5.11 Nuclear Magnetic Resonance

Solid-state NMR spectroscopy is a modern spectroscopic technique which has been widely applied in a variety of disciplines, including chemistry, biochemistry, materials science and geosciences. It is the only tool which can confirm the covalent grafting of the organic functionalities onto SNP surface. Figure 5.16(a) shows ^{29}Si CP MAS NMR of bare SiO$_2$ and APTES-grafted SiO$_2$. The ^{29}Si CP MAS NMR spectrum of bare SNSs possess three signals at -91 (Q^2), -100 (Q^3) and -109 (Q^4) ppm, which are attributed to geminal silanols, free silanols and siloxane groups, respectively. After the functionalization of SNSs with APTES, there is a reduction in the intensities of geminal (Q^2) and terminal (Q^3) silanol group signals in comparison with those of the siloxane groups. In addition to this, the broadening of the Q^4 resonance peak and appearance of a broad band around -52 ppm is observed which is assigned to T^2 and T^1 silicon environments. The ^{29}Si solid-state NMR spectrum of functionalized SNSs reveals that trialkoxysilane reagents do not exclusively form three siloxane bonds upon condensation reaction with silica surface silanol groups. The residual alkoxy groups are also evident from

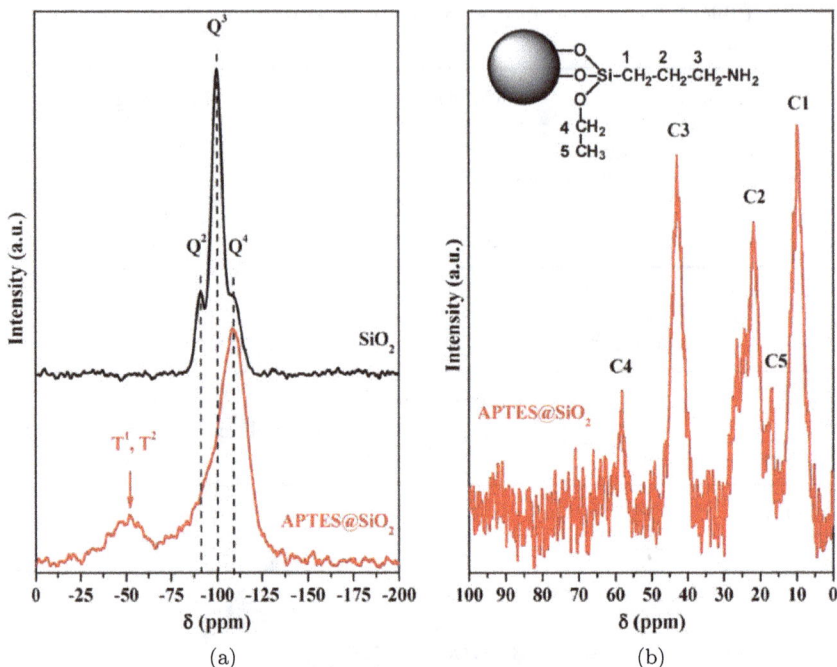

Figure 5.16. (a) ^{29}Si CP MAS NMR of SNSs and (b) ^{13}C CP MAS NMR of amine-functionalized SNSs (adapted from Ref. [21]).

the ^{13}C solid-state NMR spectrum of silica-based nanomaterials, as shown in Figure 5.16(b). The spectrum also proves the covalent bonding between the organic functionalities and the silane linking group [21].

Furthermore, the covalent immobilization of organic moieties can also be confirmed using ^{13}C CPMAS solid-state NMR spectra. The ^{13}C CPMAS NMR spectrum of SiO$_2$@APTES is shown in Figure 5.17 [8] that presents three resonance peaks at $\delta = 10.7, 23.8$ and 44.2 ppm attributed to three carbons (C1, C2 and C3, respectively) of the functionalized silane linking agent, which authenticates that the structural integrity of the APTES group is preserved even after its immobilization onto SNSs. Apart from this, the spectrum also presents two low-intensity peaks at $\delta = 16.8$ and 58.1 ppm due to

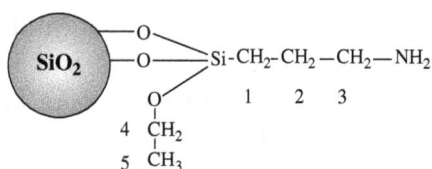

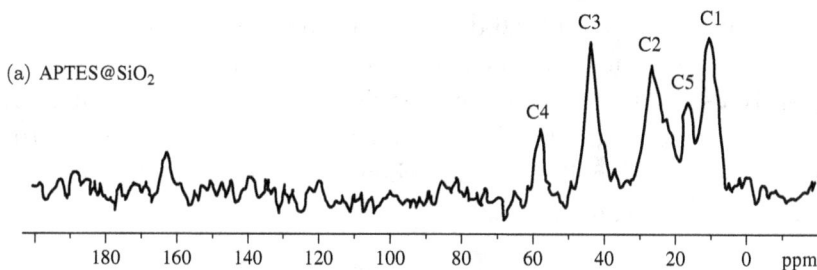

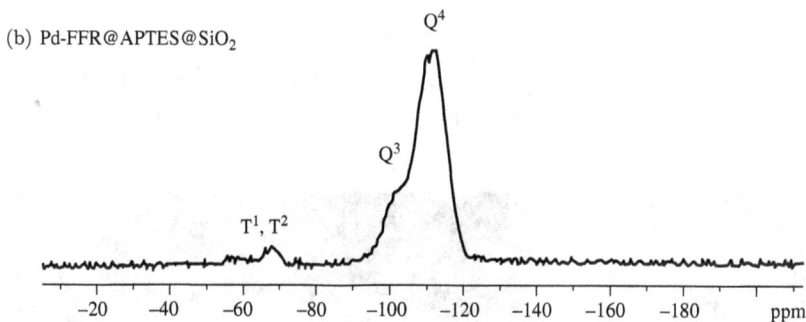

Figure 5.17. ^{13}C CPMAS NMR spectrum of SiO$_2$@APTES (adapted from Ref. [8]).

the presence of small amounts of unreacted ethoxy groups from APTES denoted as the C5 and C4 carbon atoms.

5.12 Atomic Absorption Spectroscopy/ Inductive-Coupled Plasma-Optical Emission Spectroscopy (AAS/ICP-OES)

Generally, quantitative measurement of metal content in unknown sample is determined by AAS and ICP-OES techniques. In a recent publication, the quantitative analysis of silver (Ag) content in

Ag–AcPy@ASMNPs magnetic nanocatalyst was determined by ICP-OES and the Ag content in catalyst was found to be 1.01 mmol/g [10].

Recently, Sharma and co-workers described the synthesis and application of a highly efficient, novel and magnetically retrievable solid-phase nanoadsorbent. This nanoadsorbent was prepared *via* the covalent grafting of 2,6-diacetylpyridine monothiosemicarbazide on amine-functionalized silica-encapsulated MNPs which was subsequently used for the extraction and determination of palladium (II) from different water samples and waste of catalytic solution. From AAS, amount of palladium adsorbed over the nanoadsorbent was found to be 0.63 mmol/g [22].

5.13 Scanning Transmission Electron Microscopy

Scanning transmission electron microscopy (STEM) is used to conduct the elemental mapping of an analyte sample. Elemental

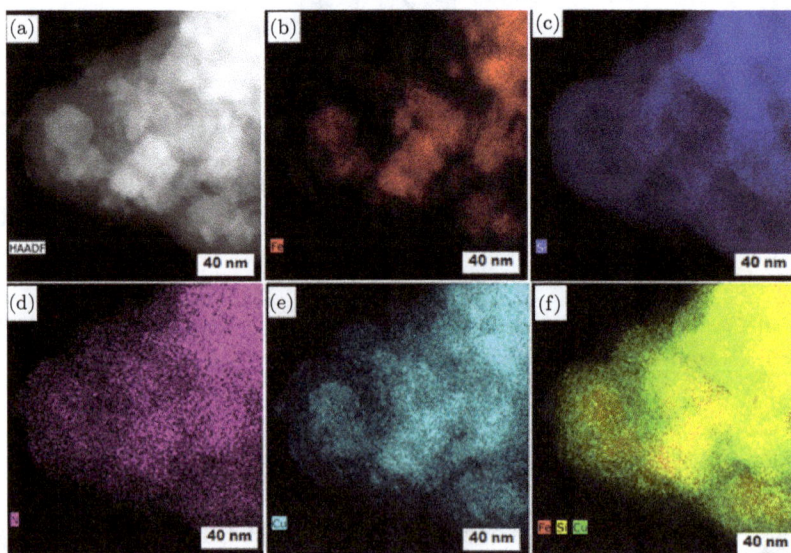

Figure 5.18. (a) High-angle annular dark-field (HAADF) image and (b)–(f) elemental mapping of Fe, Si, N, and Cu and Fe/Si/Cu of Cu–ABF@ASMNPs sample (adapted from Ref. [16]).

maps are particularly important for the distribution of elements in a sample. STEM coupled with EDS can offer more valuable insights for the core–shell structures. It provides a complete two-dimensional picture of the internal chemical zonation of the sample and elemental maps are used to utilize the sample locating spot analyses. Figure 5.18 demonstrates elemental mapping of the final Cu–ABF@ASMNPs magnetic nanocatalyst. Figure 5.18(f) clearly reveals that CuNPs are dispersed uniformly on the surface of the silica-based magnetite nanosupport. The element mapping of C cannot be determined because a huge amount of C exists in the support film of the TEM grid [16].

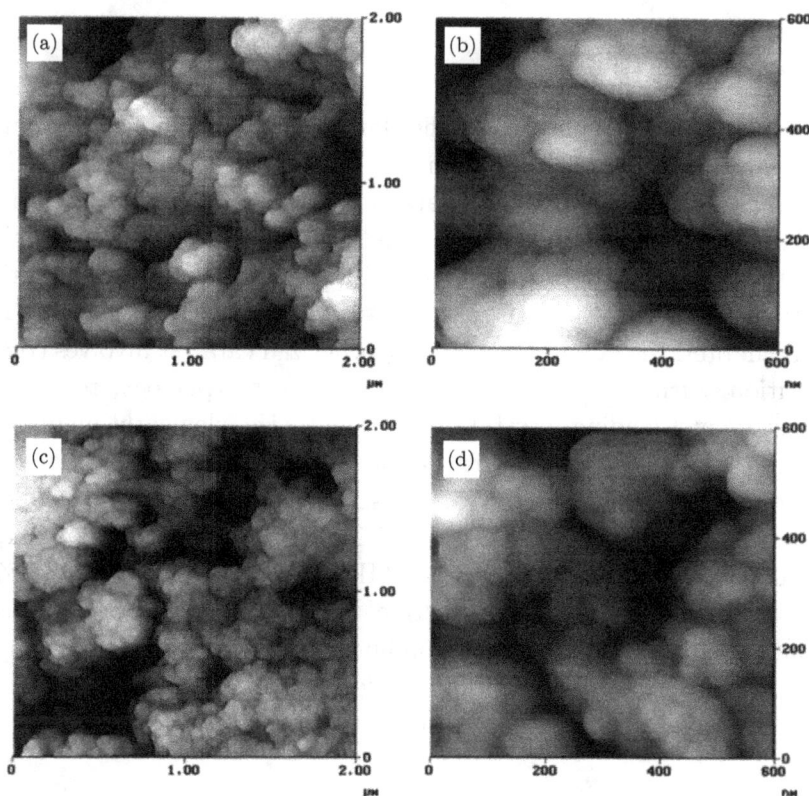

Figure 5.19.   AFM images of (a), (c) SNPs and (b), (d) polystyrene-grafted SNPs (adapted from Ref. [23]).

5.14   Atomic Force Microscopy

AFM is yet another tool for the surface characterization of the nanomaterials. This technique requires minimal sample preparation and can be used for non-conducting samples also without any specific treatment. Zhang and group have prepared nanosilica particles using a grafting polymerization method under irradiation that drastically improves the strength of a thermoplastic polymer at low filler content. The modified nanoparticles in the absence and presence of a polypropylene matrix were then characterized using AFM (Figure 5.19). The results indicated that agglomerates of untreated SNPs became more compact because of the linkage between the NPs offered by the grafting polymer [23].

5.15   Conclusion

One essential prerequisite for the development, manufacturing and commercialization of the nanomaterials is the availability of the techniques which allow the characterization of their physical and chemical properties on a nanoscale. The characterizations mainly contribute to the description and understanding of the catalytic active site (metal salt) and catalytic support (SNPs and SMNPs). In this chapter, the description of the synthesized catalyst involves composition, surface area, size, oxidation state, morphology, topology, constituents, loading, crystal arrangement, size, elemental mapping, magnetic property and identification of functional groups. Various characterization techniques are described in this chapter, which include microscopic techniques (TEM, SEM, STEM and AFM), spectroscopic techniques (XPS, FTIR, AAS/ICP-OES and EDS), scattering analysis (XRD), thermal gravimetric analysis, surface analysis (BET) and magnetization analysis (VSM). In the upcoming chapters, we will discuss the utilization of metal-immobilized SNPs and SMNPs as catalysts for various organic transformations.

References

[1] S. Chaturvedi, P. N. Dave, N. Shah, *Journal of Saudi Chemical Society* **2012**, *16*, 307–325.

[2] R. Ghosh Chaudhuri, S. Paria, *Chemical Reviews* **2011**, *112*, 2373–2433.

[3] M. Joshi, A. Bhattacharyya, S. W. Ali, *Indian J. Fibre Text. Res.* **2008**, *33*, 304–317.

[4] B. Imelik, J. C. Vedrine, *Catalyst Characterization: Physical Techniques for Solid Materials*, Springer Science & Business Media, **2013**.

[5] Y. Leng, *Materials Characterization: Introduction to Microscopic and Spectroscopic Methods*, John Wiley & Sons, **2009**.

[6] H. Fissan, S. Ristig, H. Kaminski, C. Asbach, M. Epple, *Analytical Methods* **2014**, *6*, 7324–7334.

[7] R. K. Sharma, M. Yadav, Y. Monga, R. Gaur, A. Adholeya, R. Zboril, R. S. Varma, M. B. Gawande, *ACS Sustainable Chemistry & Engineering* **2016**, *4*, 1123–1130.

[8] R. K. Sharma, S. Sharma, *Dalton Transactions* **2014**, *43*, 1292–1304.

[9] G. F. Tavares, M. R. Xavier, D. F. Neri, H. P. de Oliveira, *Chemical Engineering Journal* **2016**, *306*, 816–825.

[10] R. Gaur, M. Yadav, R. Gupta, G. Arora, P. Rana, R. K. Sharma, *ChemistrySelect* **2018**, *3*, 2502–2508.

[11] R. H.-Y. Chang, J. Jang, K. C.-W. Wu, *Green Chemistry* **2011**, *13*, 2844–2850.

[12] R. K. Sharma, Y. Monga, A. Puri, G. Gaba, *Green Chemistry* **2013**, *15*, 2800–2809.

[13] R. K. Sharma, S. Sharma, G. Gaba, *RSC Advances* **2014**, *4*, 49198–49211.

[14] R. K. Sharma, M. Yadav, R. Gaur, R. Gupta, A. Adholeya, and M. B. Gawande, *ChemPlusChem* **2016**, *81*, 1312–1319.

[15] J. Lin, W. Zhou, A. Kumbhar, J. Wiemann, J. Fang, E. Carpenter, C. O'Connor, *Journal of Solid State Chemistry* **2001**, *159*, 26–31.

[16] R. K. Sharma, R. Gaur, M. Yadav, A. Goswami, R. Zbořil, M. B. Gawande, *Scientific Reports* **2018**, *8*, 1901.

[17] F. Fajaroh, H. Setyawan, A. Nur, I. W. Lenggoro, *Advanced Powder Technology* **2013**, *24*, 507–511.

[18] M. Ghiaci, M. Zarghani, F. Moeinpour, A. Khojastehnezhad, *Applied Organometallic Chemistry* **2014**, *28*, 589–594.

[19] N. Gogoi, U. Bora, G. Borah, P. K. Gogoi, *Applied Organometallic Chemistry* **2017**, *31*, e3686.

[20] E. K. Pasandideh, B. Kakavandi, S. Nasseri, A. H. Mahvi, R. Nabizadeh, A. Esrafili, R. R. Kalantary, *Journal of Environmental Health Science and Engineering* **2016**, *14*, 21.

[21] C. Pereira, J. F. Silva, A. M. Pereira, J. P. Araujo, G. Blanco, J. M. Pintado, C. Freire, *Catalysis Science & Technology* **2011**, *1*, 784–793.

[22] R. K. Sharma, H. Kumar, A. Kumar, *RSC Advances* **2015**, *5*, 43371–43380.

[23] M. Q. Zhang, M. Z. Rong, H. M. Zeng, S. Schmitt, B. Wetzel, K. Friedrich, *Journal of Applied Polymer Science* **2001**, *80*, 2218–2227.

Chapter 6

Catalytic Applications of Silica-Based Organic–Inorganic Hybrid Nanomaterials for Different Organic Transformations

Radhika Gupta, Gunjan Arora, Manavi Yadav and
Rakesh Kumar Sharma*

*rksharmagreenchem@hotmail.com

6.1 Introduction

Catalysis holds a dynamic impact on pharmaceutical and chemical industries by facilitating the production of new materials, fine chemicals and drugs. It is the process of accelerating chemical reactions by some foreign substance known as catalyst [1]. Catalysis being a surface phenomenon is highly dependent on the nature of the surface involved. Conventionally, there are two different types of catalysts: homogeneous and heterogeneous. One of the greatest strengths of a homogeneous catalyst lies in the high accessibility of catalytically active sites which in turn presents impressive activity and selectivity. However, difficulty in separation and reusability are the major associated drawbacks that limit their large-scale industrial usability. In contrast, heterogeneous catalysts are easy to separate from reaction media but are less efficient than their homogeneous counterpart in terms of activity and selectivity [2]. Advances in this area introduced "Nanotechnology" that presents

---

*Corresponding author.

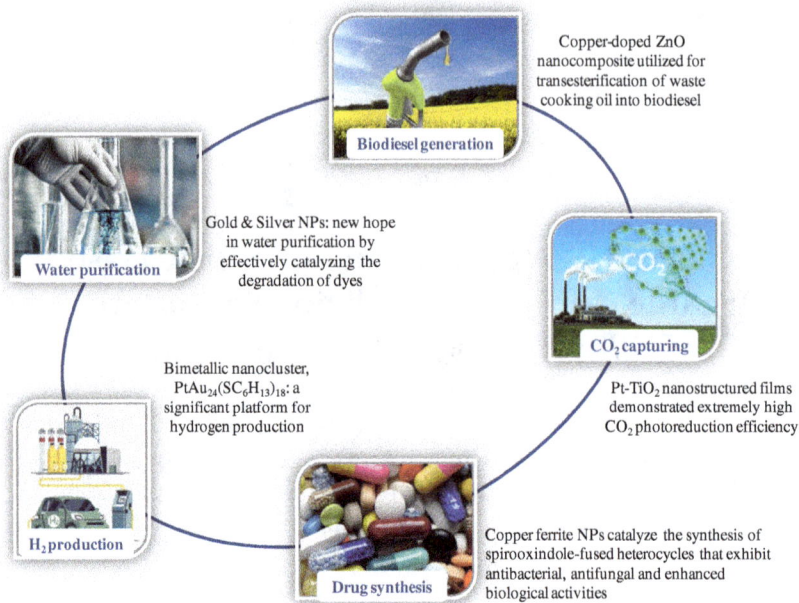

Figure 6.1. Applications of nanocatalysts.

a great opportunity for integrating the benefits of both types of catalysis [3]. It permits the design of new catalysts and catalytic processes that approach the ultimate goal: a highly active and stable catalyst that provides nearly 100% selectivity to anticipate product with minimal energy consumption. Some of the applications of nanocatalysts are mentioned in Figure 6.1 [4].

In spite of the several benefits of nanocatalysis, recovery and recyclability of the catalyst still remain crucial points of interest which hamper the large-scale industrial application of such processes. In order to combat these shortcomings, environment-friendly and sustainable supported nanocatalysis has gained tremendous attention [5]. Over the last few years, considerable efforts have been made for the design and fabrication of silica-based nanostructures [6]. Due to their unique nano-size properties, these materials have shown widespread applications in the field of biomedicine for diagnosing and controlling diseases [7], identifying and correcting genetic disorders, biosensing [8] and other therapeutic applications, such as photodynamic therapy [9]. Also, silica-based magnetic nanoparticles (SMNPs) are immensely attractive from the drug delivery viewpoint because they can respond to external stimuli *via* a magnetic field [10].

In this chapter, we have discussed silica/magnetic silica-coated core−shell nanocomposites as ideal supports for catalyst immobilization. Owing to the remarkable properties of these nanostructures, the catalytic activity of the immobilized systems can be greatly improved in terms of mild reaction conditions, better yield and selectivity of the desired product, easy recovery from the reaction media and hence, recyclability of the nanocatalyst. Besides, the binding of catalytic species also prevents the product stream from getting contaminated with the catalyst residues. This will prevent the necessity of additional purification methods to remove the catalyst from the reaction media making them "greener" in comparison to previously reported approaches (Figure 6.2) [11].

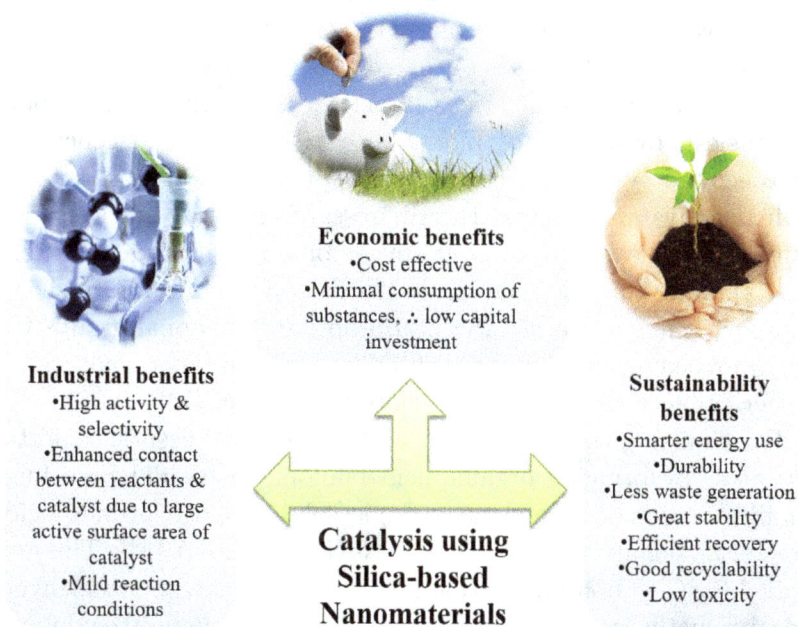

Economic benefits
•Cost effective
•Minimal consumption of substances, ∴ low capital investment

Industrial benefits
•High activity & selectivity
•Enhanced contact between reactants & catalyst due to large active surface area of catalyst
•Mild reaction conditions

Sustainability benefits
•Smarter energy use
•Durability
•Less waste generation
•Great stability
•Efficient recovery
•Good recyclability
•Low toxicity

Catalysis using Silica-based Nanomaterials

Figure 6.2. Benefits of catalysis using silica-based nanomaterials.

Herein, we describe the use of numerous silica-based metal-/acid-immobilized organic–inorganic hybrid nanomaterials which have been used for catalyzing numerous organic reactions, such as coupling, oxidation, reduction, multicomponent, condensation, CO_2 capturing and many more.

6.2 Applications of Silica-Based Nanomaterials in Catalysis

6.2.1 Coupling reactions

6.2.1.1 *C–C coupling*

Systematic development of a multifunctional catalyst in complex organic transformation can serve the purpose of controlling a multistep process. This is usually observed in the case of biological catalysts, where precise positioning of active centers accelerate the rate and control the activity and selectivity (through acid–base, electrostatic interactions, hydrogen bonding or covalent bonding) of the reactive species. However, one of the biggest challenges in the field of heterogeneous catalysts is to control the spatial arrangement along with the relative concentrations of active site on the solid support. Nonetheless, by mimicking biological catalysts, one can generate a heterogenized system wherein the concept of site isolation can lead to dual activation. In this regard, researchers have functionalized multiple organic groups on uniform mesoporous silica nanoparticles (MSNs) under suitable conditions [12]. Due to the presence of silanol groups on the surface of MSNs, appropriate functional groups can be introduced inside pore channels and additional functionalities can further enhance the activity *via* cooperative effect [13].

On similar grounds, Shylesh and co-workers recently reported a catalyst by bringing organic acid (sulfonic acid) and base functionalities (amino groups) on a single solid support forming acid–base bifunctional nanocatalyst (MSN–NNH$_2$–SO$_3$H) [14]. This was applied for C–C bond formation between substituted benzaldehydes and nitromethane along with one-pot deacetalization-nitroaldol and one-pot deacetalization-aldol reactions (Scheme 6.1). Unlike the homogeneous system, where the problem of coexistence of acid–base

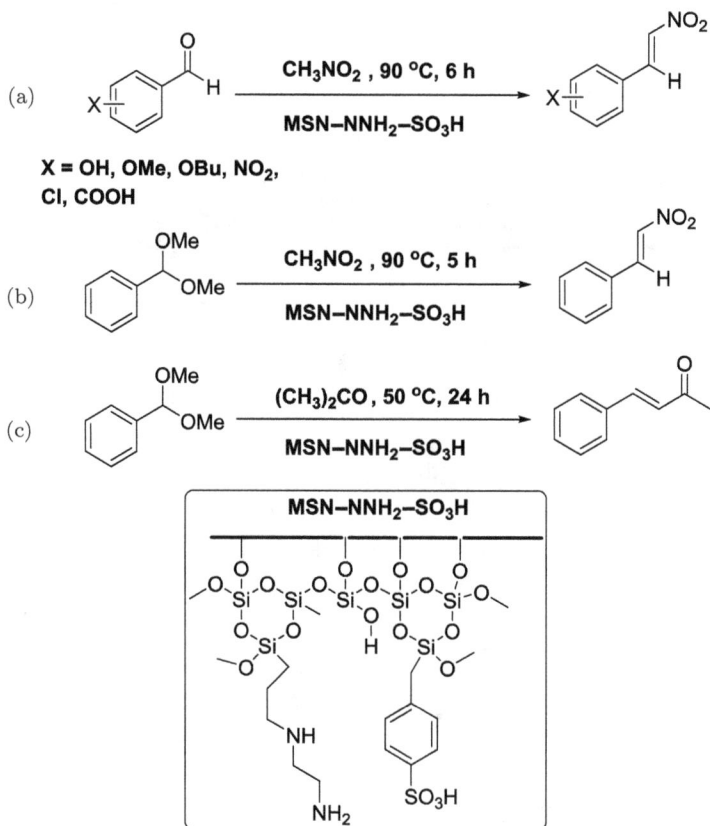

Scheme 6.1. MSN–NNH$_2$–SO$_3$H-catalyzed (a) nitroaldol, (b) deacetalization-nitroaldol and (c) deacetalization-aldol reaction.

occurred due to the mutual neutralization, synergistic catalytic enhancements were observed in the heterogenized system due to the cooperative catalytic reactions and controlled spatial distribution on silica support.

Another useful C–C bond forming reaction includes Suzuki–Miyaura coupling which is well documented in numerous research articles [15]. Various transition metal-based catalysts have been reported till date to catalyze these coupling reactions, Pd metal being the first choice among all. In this regard, various research groups have fabricated palladium-supported silica-coated magnetic

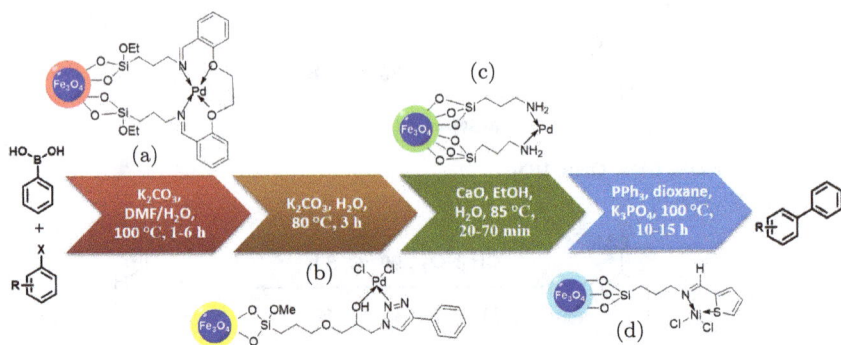

Scheme 6.2.   Various metal-based silica-coated magnetic nanocatalysts for Suzuki reaction.

nanocatalysts for the coupling reaction (Scheme 6.2). Azadbakht *et al.* synthesized a water-stable nanocatalyst by immobilizing (2-[2-(2-formylphenoxy)ethoxy]benzaldehyde) dihydroxyaryl palladium complex on amine-functionalized SMNPs (Scheme 6.2(a)) [16]. The developed catalyst was utilized in the coupling of phenylboronic acids and aryl halides. The cross-coupling products were obtained in high yields under ambient atmosphere in aqueous medium and in short reaction time.

Another group designed Pd@Fe$_3$O$_4$ nanocatalyst to catalyze the coupling of aryl halides and especially inactive aryl chlorides with benzenoidaromatic rings (Scheme 6.2(b)) [17]. The catalyst circumvents the use of organic solvents and toxic phosphine ligands. Short reaction times, high yields and purity of the products and stability of the catalyst toward air, moisture and heat are some of the other advantages of the reported catalyst.

In order to develop a greener catalyst for Suzuki reaction, Khazaei *et al.* designed nano-Fe$_3$O$_4$@SiO$_2$-supported Pd(0) (Fe$_3$O$_4$@SiO$_2$–Pd) in which rice husk biomass was used for obtaining biosilica (Scheme 6.2(c)) [18]. Also, in the coupling reaction, waste eggshell was used as a low-cost solid base. Thus, the presented methodology provides significant advantages, such as use of natural and waste materials and recyclability of the catalyst.

While considering the limited reserves and extinction of palladium metal in the coming future, Sharma and co-workers developed

$SiO_2@Fe_3O_4$-based nickel base metal catalyst for the formation of biaryls (Scheme 6.2(d)) [19]. Wide substrate scope, mild reaction conditions, high yields and reusability of the catalyst up to six consecutive cycles are some of the key features that are beneficial from the industrial viewpoint.

Similarly, Hajipour and Azizi reported the synthesis of aromatic ketones by introducing acyl halides in place of aryl halides in the Suzuki–Miyaura reaction using $Fe_3O_4@nSiO_2@pSiO_2–Pd(acac)_2$ as the catalyst (Scheme 6.3) [20]. The above catalyst was synthesized by supporting palladium (II) acetate on acetylacetonate functionalized double SMNPs. All the ketones were obtained in good to excellent yields in just one step and under anhydrous reaction conditions.

Besides Suzuki reaction, SMNPs-based catalysts have also been utilized for the oxidative α-cyanation of tertiary amines. A report by Yang *et al.* described the first example of the use of heterogeneous magnetic nanoparticles (MNPs)-supported gold(III)-bipy complex as catalyst for the oxidative α-cyanation of tertiary amines (Scheme 6.4) [21]. The electronic effect of substituents has limited impact on this heterogeneous gold-catalyzed oxidative α-cyanation reaction, and a variety of functional groups including

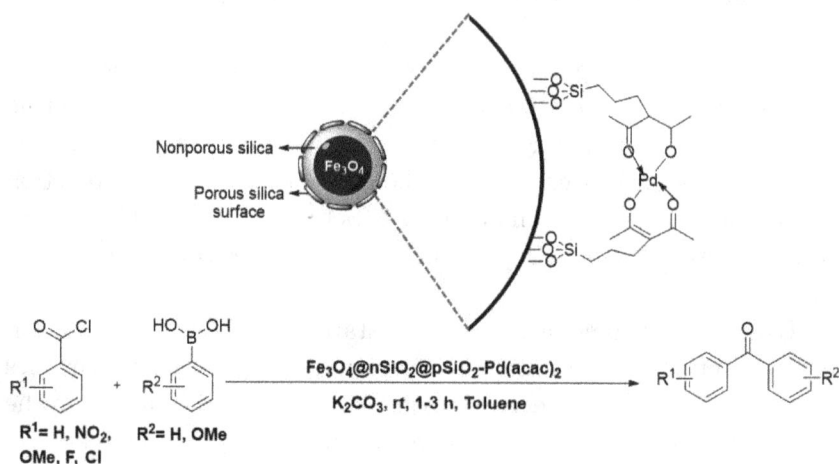

Scheme 6.3. Synthesis of aromatic ketones using $Fe_3O_4@nSiO_2@pSiO_2–Pd(acac)_2$.

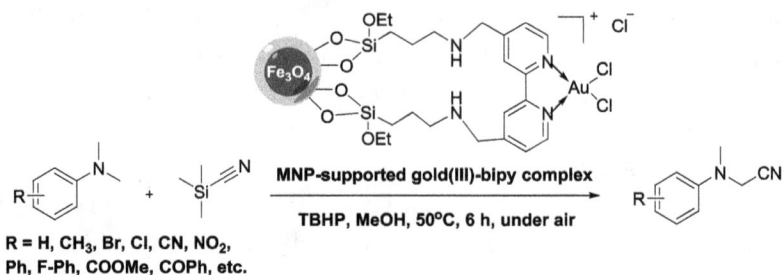

Scheme 6.4. Oxidative α-cyanation of tertiary amines using MNP-supported gold(III)-bipy complex.

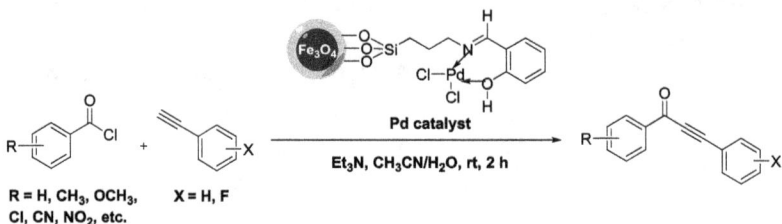

Scheme 6.5. Sonogashira coupling reaction of acyl chlorides and terminal alkynes.

both electron-donating and -withdrawing groups were selectively transformed to their corresponding α-aminonitriles in good to excellent yields.

Sonogashira coupling is one of the powerful transformations for the formation of $C(sp^2)$–$C(sp)$ bond [22]. Thus, considering the widespread applications of ynones, palladium-based magnetic nanocatalysts have been used for the Sonogashira coupling reaction of acyl chlorides and terminal alkynes (Scheme 6.5) [23]. The ynones were synthesized under copper- and phosphine-free conditions and in excellent yields.

Usually, bare nanoparticles are unstable and their agglomeration during reaction is inevitable. Till now, studies on native nanoparticles as catalysts are limited, even though they have significant role in the field of catalysis. Recently, the scope of silica NPs (SNPs) as catalyst for anti-Markovnikov addition of thiols to alkenes and alkynes was explored [24]. Michael addition is considered one of the advantageous C–C bond-forming reactions and is simply catalyzed by strong bases

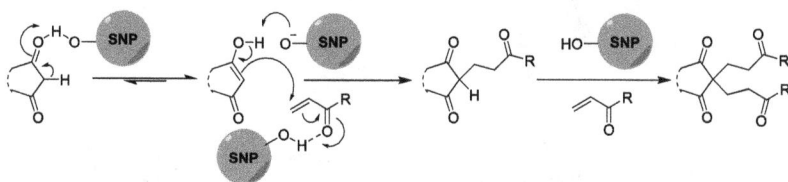

Scheme 6.6. Probable mechanism of SNP-catalyzed bis-Michael addition.

which often lead to undesired products. Although a number of reagents have been reported so far for the mono-Michael addition reaction, most of them failed to trigger the bis-Michael addition of active methylene compounds to conjugated ketones, carboxylic esters and nitriles, even at higher temperature conditions. Banerjee and Santra evaluated the catalytic activity of native SNPs in an unusual bis-Michael addition and observed phenomenal results under neutral conditions with no mono-addition product [25]. The bis-Michael addition of active methylene compounds to α, β-unsaturated ketones, esters and nitriles was achieved in a single step.

While this exceptional behavior is unknown, it is believed that the surface chemistry of silica plays a significant role in this reaction (Scheme 6.6). Since at neutral pH, SNPs provide both –Si–OH and –Si–O$^-$ groups on its surface, the –Si–OH group stabilizes the enol form of active methylene compounds and forms H-bonds with carbonyl oxygen thereby polarizing the conjugated alkene, whereas Si–O$^-$ stimulates the nucleophilic attack of the enolized active methylene compound.

6.2.1.2 *C–N coupling*

C–N bond formation reactions are considered to have paramount importance as these are applicable for the synthesis of many industrially important products [26]. In view of this, formation of C–N bonds with readily available substrates has garnered tremendous attention. Recently, Inamdar *et al.* explored a green and efficient method for hydroacylation reaction of various aldehydes with diisopropyl azodicarboxylate, using silica-supported copper nanocatalyst (CuO–NP/SiO$_2$) where azodicarboxylates are used as electrophiles (Scheme 6.7) [27]. The high efficiency of this reaction is

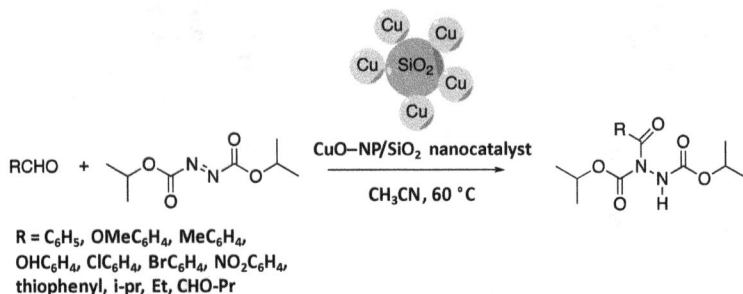

RCHO + [structure] →(CuO–NP/SiO₂ nanocatalyst, CH₃CN, 60 °C)→ [product structure]

R = C₆H₅, OMeC₆H₄, MeC₆H₄,
OHC₆H₄, ClC₆H₄, BrC₆H₄, NO₂C₆H₄,
thiophenyl, i-pr, Et, CHO-Pr

Scheme 6.7. CuO–NP/SiO₂-catalyzed hydroacylation reaction of aldehydes and azodicarboxylate.

subjected to the electron-withdrawing nature of azodicarboxylates. The other noticeable features of this protocol involve inexpensive and recyclable nanocatalyst, excellent product yield, ligand- and additive-free reaction and method simplicity.

Among C–N coupled compounds, alkyl-substituted amines have gained commercial importance due to their usage in fine chemicals and pharmaceutical industries [28]. On this note, Sharma *et al.* have fabricated a highly selective magnetic nanocatalyst by immobilizing copper on 2-acetylthiophene-functionalized SMNPs (Cu–AcTp@Am–Si–Fe₃O₄) (Scheme 6.8) [29]. For this, alcohols were chosen as green alkylating agents. Mild reaction conditions, zero effluent discharge, broad substrate scope, high product yield and recyclability of the catalyst are some of the features which make the protocol superior from the previously reported methodologies.

To make the procedure of catalyst preparation easier and more convenient, Baig and Varma developed a one-step protocol for the synthesis of CuSO₄-supported SMNPs (Fe₃O₄@SiO₂Cu) [30]. The catalytic activity of the synthesized catalyst was checked in the amination of aryl halides (Scheme 6.9). Results showed that all the arylated amines were obtained in very good yields.

6.2.1.3 C–O coupling

Diaryl ethers are one of the most important C–O coupling products due to their widespread applications in pharmaceuticals, polymers and agrochemicals [31]. They also appear in biologically active

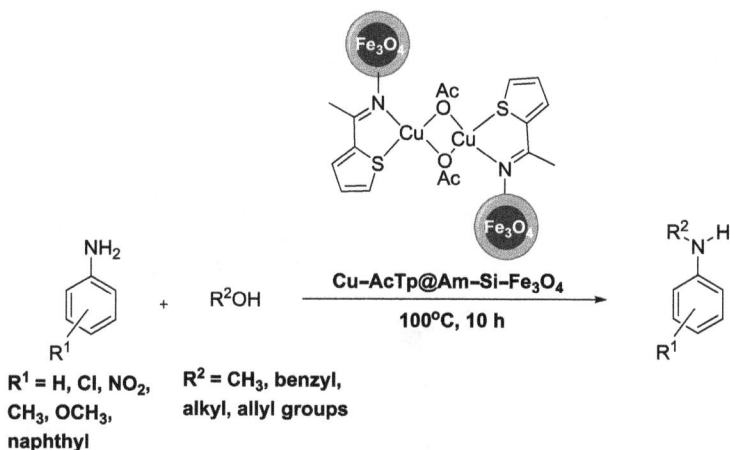

Scheme 6.8. Synthesis of alkyl-substituted amines using Cu–AcTp@Am–Si–Fe_3O_4.

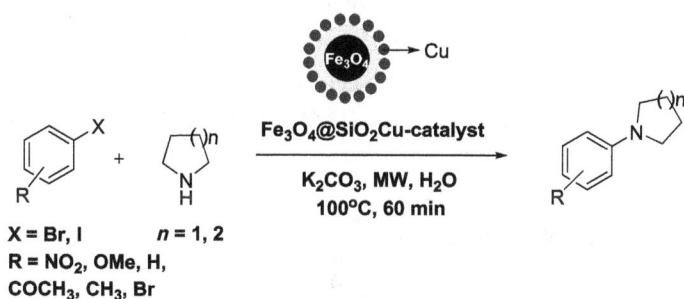

Scheme 6.9. Amination of aryl halides using $CuSO_4$-supported SMNPs.

natural products. Therefore, significant amount of research has been done for their synthesis. Zolfigol *et al.* have designed water-stable Pd-containing phosphorus silica-based magnetic nanocatalyst (Fe_3O_4@SiO_2@PPh_2@Pd(0)) for the aqueous-phase *O*-arylation of phenols with aryl halides (Scheme 6.10) [32]. The protocol provides a cleaner reaction profile, good product yields and recyclability of the catalyst.

6.2.1.4 *C–S coupling*

C_{aryl}–S bond-forming reactions are comparatively less studied because of the tendency of thiols to undergo self-oxidation, and

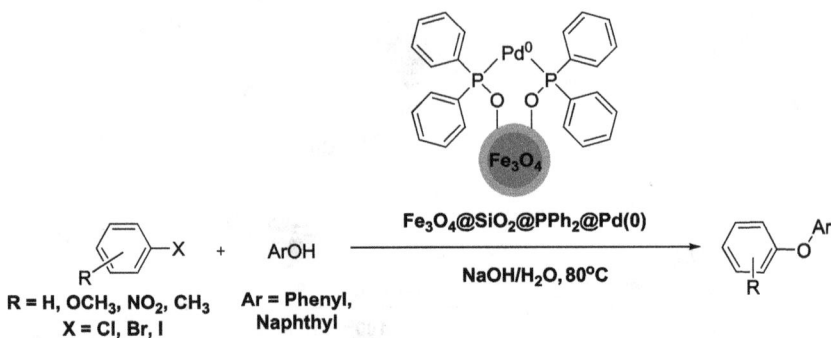

Scheme 6.10. *O*-arylation of phenols with aryl halides.

thereby forming disulfides. Keeping the above drawback in mind and with the aim of coupling aryl halides with thiols, Movassagh and co-workers recently utilized the chelation properties of azacrown ether to form stable complexes with metal ions [33]. They fabricated heat- and air-stable SMNP-supported palladium(II)-cryptand 22 complex [Fe_3O_4@SiO_2@C22–Pd(II)] as a new heterogeneous catalyst for forming C_{aryl}–S and C_{aryl}–C_{aryl} coupling reactions (Scheme 6.11). All the coupling products were formed in good to excellent yields.

6.2.2 Oxidation reactions

Silica-based metal nanocatalysts have also been used for various oxidation reactions (Table 6.1). Since oxidation products are useful for the manufacturing of agrochemical and pharmaceutical compounds [34], many reactions have been reported for their formation using traditional oxidants such as chromic acid and selenium dioxide, but formation of hazardous and corrosive wastes and tedious workup procedures limit the green reaction credentials [35]. In this regard, Chen *et al.* have developed a copper-based nanocatalyst (Fe_3O_4/SiO_2/Cu(II)salpr) for the selective oxidation of alkyl aromatic compounds into corresponding ketones using *tert*-butyl hydroperoxide (TBHP) as the oxidant under solventless conditions (Table 6.1, Entry 1) [36]. TBHP is considered as an environmentally benign oxidant since only *tert*-butanol is formed as a by-product. Results showed that the oxidation reaction took place on α-C of the

Fe₃O₄@SiO₂@C22–Pd(II)

(a)

X

R^1———————— + R^2SH $\xrightarrow[\text{KOH, DMSO, 80°C}]{\text{Catalyst}}$ R^1————————

SR²

X = Cl, Br, I   R² = Ph, C₆H₄OMe,
R¹ = H, Me, OMe   hexyl

(b)

X   HO   OH
                       B

R^1———— + R^2———— $\xrightarrow[\text{DMF/H}_2\text{O, 75°C}]{\text{Catalyst, Et}_3\text{N}}$ R^1————————R^2

X = Cl, Br, I   R² = H, OMe
R¹ = H, Br, CN,
Me, OMe, Me,
NO₂, CHO

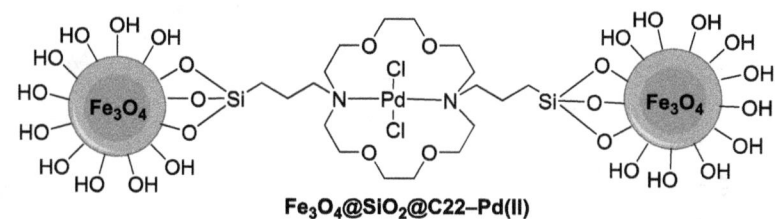

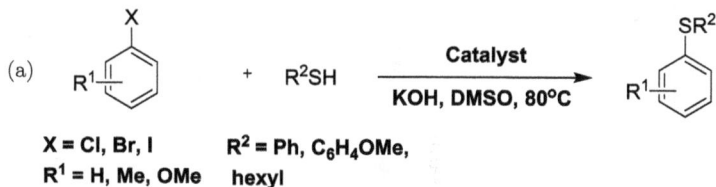

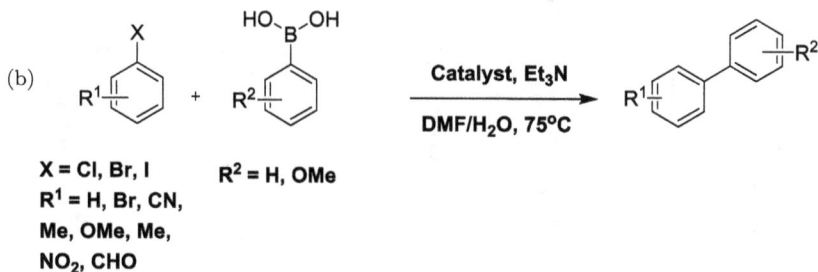

Scheme 6.11. (a) C_{aryl}–S and (b) C_{aryl}–C_{aryl} coupling reactions using Fe₃O₄@SiO₂@C22–Pd(II).

alkyl aromatic compounds and thereby forming ketones as the major product. Only minor quantities of the corresponding aldehyde and acid were obtained when the reaction was carried out at 80°C and for 12 h. Besides providing selectivity, the catalyst offers other useful advantages, such as easy magnetic recovery and reusability.

Also, metalloporphyrin complexes have been widely used as biorelevant catalysts in oxidation chemistry [37]. Immobilization of these complexes minimizes self-destruction of catalysts and dimerization of unhindered metalloporphyrins [38]. On this note, Rezaeifard *et al.* have immobilized manganese (III) acetate meso-tetraphenylporphyrinato complex onto SMNPs (Mn(TPP)OAc@ SMNP) and used it for the epoxidation of olefins and for the oxidation

Table 6.1. Various silica-based nanocatalysts that have been utilized for different oxidation reactions.

S.No.	Reactant	Oxidized product	Catalyst	Oxidant and reaction conditions
1.	R = linear alkyl chains, benzyl, fused cyclic hydrocarbons X = H, OMe, NO$_2$, Cl, Br		**Fe$_3$O$_4$/SiO$_2$/Cu(II)salpr**	80% TBHP, 80°C, 12 h, solventless
2.				TBHP, 60°C, 4.5–12 h, water, under air
	R = linear alkyl chains, fused cyclic hydrocarbons, substituted hydroxy alkyl chains X = H, OMe, NO$_2$, Cl		**Mn(TPP)OAc@SMNP**	TBHP, 60°C, 4.5 h, water, under air
				TBHP, 25°C, 1 h, water, under air

#	Substrate	Product	Catalyst	Conditions
3.	X = Cl, Br R = H, Me, OMe, NO_2, Cl, Br		**Mn–Ac@ASMNP**	H_2O_2, ethanol, reflux, 1.5–2.5 h
4.	R = H, Cl, Br		**Zn(II)Ac–Py@ASMNP**	H_2O_2, solventless, reflux, 0.5–1 h
	R = CH_3, OCH_3, Cl, Br, NO_2, $COCH_3$			H_2O_2, 80°C, 1 h, acetonitrile
5.	Cyclohexene	Cyclohexene epoxide, 2-cyclohexen-1-ol, 2-cyclohexen-1-one	**MSS–SH–Au⁰**	O_2, 100°C, 8 h
	Styrene	Styrene epoxide, benzaldehyde, acetophenone		

(Continued)

Table 6.1. (*Continued*)

S.No.	Reactant	Oxidized product	Catalyst	Oxidant and reaction conditions
6.	styrene	phenyloxirane	M= Cu or Co $Fe_3O_4@SiO_2-NH_2-M$	Air, 80°C, 8 h, CH_3CN
7.	R-SH R = Ph, Ph-CH_3, Ph-OCH_3, Ph-NH_2, Cl, Br, hexyl, propyl	R–S–S–R	NO_3Ag Ag–AcPy@ASMNP	Air, 30°C, 30 min, water
8.	X—(benzyl alcohol)—OH X = H, Me, OMe, Cl, Br, NO_2	X—(benzaldehyde)	Ag NPs $Fe_3O_4@SiO_2-Ag$	Refluxing with toluene, 24 h, N_2 atmosphere

of hydrocarbons, alcohols and sulfides into the corresponding ketones, aldehydes and sulfoxides in excellent conversion and selectivity percentages (Table 6.1, Entry 2) [39]. Since a very low amount of catalyst was used for carrying out the reaction in water and using TBHP, the dispersibility increases in aqueous solution which might be a reason for such a high catalytic activity.

Sharma *et al.* have recently developed a manganese-based magnetically retrievable nanocatalyst by immobilizing manganese (III) acetylacetonate complex on functionalized SMNPs (Mn–Ac@ ASMNP) [40]. This nanocatalytic system was efficiently utilized for the oxidation of benzyl halides and alcohols using the sole and green oxidant H_2O_2 (Table 6.1, Entry 3).

Using the same oxidant, H_2O_2, various aromatic amines were oxidized to azoxyarenes by making use of zinc-supported magnetic nanocatalyst (Ag–AcPy@ASMNP) (Table 6.1, Entry 4) [41]. The reported methodology is highly desirable from the industrial viewpoint as all the oxidation products were obtained in high yields and selectivity.

Even though hydrogen peroxide is considered as a green oxidant, it is derived from synthesis of gas (CO and H_2) [42]. This has driven the researchers to explore the reactivity of late-transition metal complexes toward molecular oxygen [43]. Such oxidants are clean, green, safe and economic oxidants for various reactions. A group oxidized cyclohexene and styrene using molecular oxygen at atmospheric pressure by making use of gold supported on thiol-modified silica coated magnetic mesoporous nanocrystals (MSS-SH-Au^0) (Table 6.1, Entry 5) [44]. The mesoporous structure of the silica support provides large surface-active sites which helps in the easy grafting of metal affinity groups, such as $-SH$.

Another group has developed Fe_3O_4@SiO_2-supported cobalt(II) and copper(II) acetylacetonate complexes (Fe_3O_4@SiO_2–NH_2-M) for epoxidation of styrene using only air as the oxygen source (Table 6.1, Entry 6) [45]. Both of the metal nanocatalysts showed excellent styrene conversion and good selectivity for the epoxide formation (conversion % with Co = 90.8% and with Cu = 86.7%; selectivity % with Co = 63.7% and with Cu = 51.4%). The probable explanation for the difference in catalytic activity is due to the better performance of Co ions to activate the molecular oxygen. The developed heterogeneous catalysts also showed better catalytic activities than their homogeneous counterparts. This is possibly due to the fact that heterogenization of transition metal complexes can effectively prevent the dimerization of active sites and also reduce the formation of μ-oxo, μ-peroxo dimeric and other polymeric species, which usually occur in the homogeneous catalytic systems. Besides synergistic effects between metal complexes and silica-coated nanosupport, high loading of active sites due to the large surface area-to-volume ratio of nanocatalyst also significantly increases the activity of the heterogeneous catalyst.

Similarly, Sharma and co-workers synthesized silver-based magnetic nanocatalyst (Ag–AcPy@ASMNP) for the oxidation of thiols into disulfides (Table 6.1, Entry 7) [46]. The oxidation reaction was carried out at room temperature within 30 min by simple stirring in water and in the presence of air.

Bayat *et al.* fabricated SMNPs-based nanocatalyst by depositing Ag nanoparticles on SMNPs (Fe_3O_4@SiO_2–Ag) [47]. The catalytic activity of the synthesized catalyst was tested for the oxidant-free dehydrogenation of alcohols (Table 6.1, Entry 8). Under the optimized reaction conditions, a variety of substituted alcohols were converted into their corresponding carbonyl compounds with high chemoselectivity and in excellent yields.

Another work was reported, where gold nanoparticles (Au NPs) with varied average size (ranging from 1–13 nm) were immobilized on fibrous silica nanospheres (SNSs) (KCC-1) either by binding pre-made Au NPs on KCC-1-NH_2 or by tethering $HAuCl_4$ on KCC-1-NH_2 followed by reduction (Figure 6.3) [48]. With the latter method, highly dispersed and uniform Au NPs were located inside the fibers of KCC-1-NH_2, giving an average size within 5 nm with high metal loading of about 10.4 wt.%. The resulting nanostructures did not display any sign of agglomeration even when they were subjected to H_2 reduction at 300°C for 1 h. These were then applied in catalysis for CO oxidation at temperatures ranging from 100 to 300°C. It was observed that the catalytic activity of the

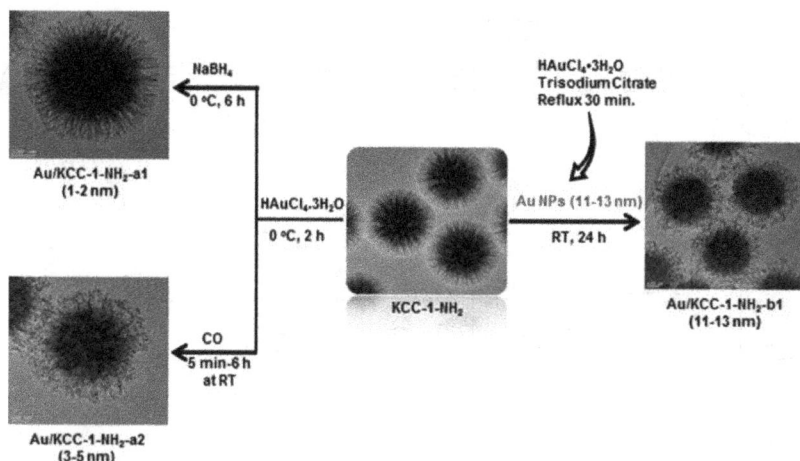

Figure 6.3. Synthesis of Au/KCC-1-NH$_2$ (adapted from Ref. [48]).

Au NPs was size-dependent. Nonetheless, the performance did not vary much between the catalysts Au/KCC-1-NH$_2$-a1 (1–2 nm size, Au loading 5.6 wt.%) and Au/KCC-1-NH$_2$-a2 (3–5 nm size, Au loading 10.4 wt.%). However, the latter catalyst was found to be more stable toward aggregation than the former upon CO treatment. Also, Au/KCC-1-NH$_2$-b1 (11–13 nm, Au loading 5.8 wt.%) displayed reduced catalytic performance as they could not penetrate the pores of KCC-1 due to their large size. Thus, it was concluded that the region and size of Au NPs on the support depend on the synthetic procedure adopted. This method demonstrated a suitable preparation method for synthesizing highly dispersed Au/KCC-1 nanocatalysts with varied metal loadings and good catalytic activities.

Another Pt-loaded MSN was developed using two different approaches and applied for CO oxidation (Scheme 6.12) [49]. In the first approach, Pt was immobilized on silica support by reduction of Pt-containing salt, and this resulted in a triple-nanostructural catalytic material. It was found that at a low ratio of Pt and SiO$_2$, the catalytic activity toward CO oxidation by mesoporous silica support is much higher than that with micropores, while the second approach combined the synthesis of monolithic mesoporous silica with Pt loading. This was considered to be more convenient and

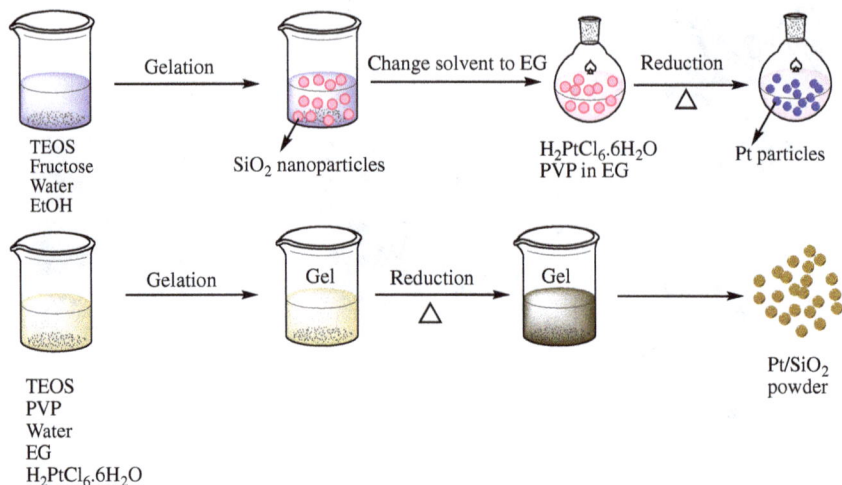

Scheme 6.12. Schematic illustration of the synthetic processes of Pt/SiO$_2$ catalysts in two different ways.

Note: PVP = Polyvinylpyrrolidone; EG = Ethylene glycol.

economical as it was prepared in a single step and the template for mesoporous silica was also utilized in reducing Pt. This also displayed higher activity for CO oxidation than that with micropores.

Decomposition of volatile organic compounds (VOCs) is an extremely important process since they are considered as major air pollutants released by industries [50]. More than 300 compounds have been categorized as VOCs including aromatic hydrocarbons, oxygenates and halohydrocarbons. A suitable method for its removal is catalytic oxidation [51], but advanced processes include a combination of adsorption unit and catalytic incinerator. Recently, Popova and co-workers designed a system composed of both adsorption and catalytic units, by utilizing transition metal oxides instead of the presently employed noble metal-containing catalysts [52]. The group demonstrated efficient adsorption and catalytic activity of iron-functionalized silica nanoparticles (SNPs) with interparticle mesoporosity in toluene oxidation in gas phase. It was also reported through spectroscopic studies that the superior catalytic activity of the developed matrix was attributed to the formation of stable Fe^{3+} ions in the silica matrix which could release oxygen easily through

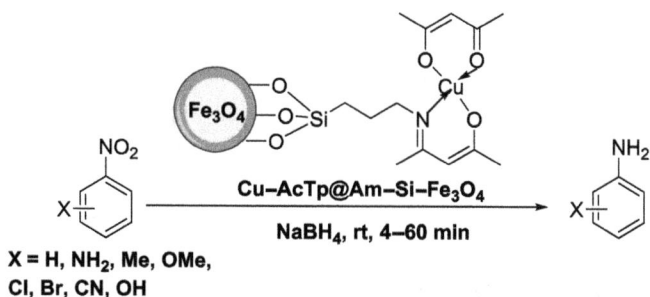

Scheme 6.13. Reduction of nitroarenes using Cu–AcTp@Am–Si–Fe$_3$O$_4$.

Fe^{3+}/Fe^{2+} redox cycles. Thus, the material possesses promising features including having low-cost, being environmentally friendly and possessing high efficiency for the removal of low-concentration VOC from polluted air.

6.2.3 Reduction reactions

Reduction of aromatic nitro compounds has garnered considerable amount of attention since decades due to their widespread application in dye synthesis, agrochemicals, pharmaceuticals and other fine chemical industries [53]. Famous traditional reduction routes involve catalytic hydrogenation using metal-based catalysts, such as Ru, Pd, Pt, Bi, Pt/Ni, Pt/Pd and V. But these methodologies suffer from numerous disadvantages such as limited metal reservoirs and economic constraints due to their high cost [54]. In this regard, Sharma and co-workers reported the fabrication and characterization of silica@Fe$_3$O$_4$-based heterogeneous copper catalyst (Cu–AcTp@Am–Si–Fe$_3$O$_4$) for the reduction of nitroarenes using NaBH$_4$ at room temperature in aqueous medium (Scheme 6.13) [55]. The catalyst was fabricated by grafting copper (II) acetylacetonate complex on amine-functionalized SMNPs. Using this catalyst, a number of substituted aromatic nitro compounds were reduced into the corresponding amines with excellent conversion and selectivity percentages.

The same research group also fabricated a similar magnetic nickel nanocatalyst by covalently immobilizing 2-acetyl furan on the surface

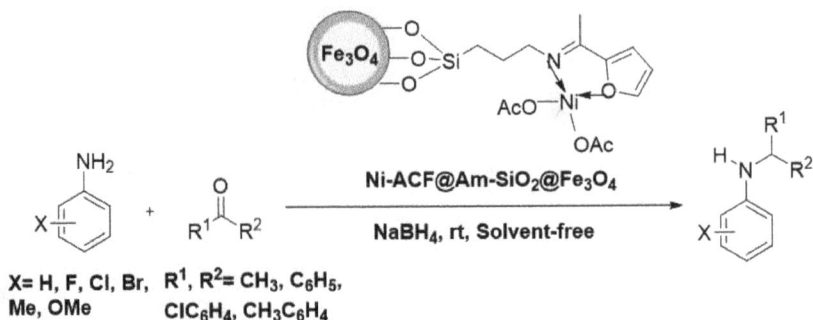

Scheme 6.14. One-pot reductive amination of ketones.

of amine-functionalized SMNPs and further metallating with nickel acetate [56]. The catalytic performance of the above catalyst was investigated in the direct one-pot reductive amination of ketones using $NaBH_4$ under solvent-free conditions and at room temperature (Scheme 6.14). All the secondary amines were synthesized in good yields.

6.2.4 Multicomponent reactions

Multicomponent reactions (MCRs) correspond to one of the most efficient synthetic protocols in which more than two reactants combine together in a single reaction vessel to form a highly selective product that retains a majority of the atoms of the starting materials. MCRs have all the features for ideal synthesis: high atom economy, time and energy saving, almost no by-product and one-pot operation procedure [57]. Therefore, MCRs are gaining much attention and a diverse array of silica-based nanocatalysts has been reported in literature for these reactions (Table 6.2). Khojastehnezhad *et al.* have synthesized ferric hydrogen sulfate-modified silica-coated nickel ferrite nanoparticles ($NiFe_2O_4$@SiO_2–FHS) [58]. The so-formed catalyst was efficiently utilized for carrying out the MCR of aromatic aldehyde, dimedone and ammonium acetate/aromatic amine to form 1,8-dioxodecahydroacridines under solvent-free conditions (Table 6.2, Entry 1).

Table 6.2. Silica-based nanocatalysts for various MCRs.

S.No.	Catalyst	Reactants	Product	Conditions	Yield (%)	No. of examples
1.	NiFe$_2$O$_4$@SiO$_2$–FHS	NH$_4$OAc or RNH$_2$ R = C$_6$H$_5$, 4-MeC$_6$H$_4$, 4-MeOC$_6$H$_4$ R^1 = H, Me, OMe, Cl, Br, NO$_2$	1,8-dioxodecahydro-acridine derivatives	NiFe$_2$O$_4$@SiO$_2$–FHS(1 mol%), 80°C, solvent-free	82–94	14
2.	[MNPs@GLU][Cl]	R^1 ≡ + R^2 ⌐ Br NaN$_3$ R^1, R^2 = alkyl/aryl	1,3,5-triazole derivatives	[MNPs@GLU][Cl] (0.5 mol%), 50°C, water	85–99	22

(*Continued*)

Table 6.2.   (*Continued*)

S.No.	Catalyst	Reactants	Product	Conditions	Yield (%)	No. of examples
3.	SiO₂ SiO₂NPs	RCHO + $\begin{smallmatrix}CN\\CN\end{smallmatrix}$ + dimedone R = Et, Pr, aryl derivatives	4H-pyran derivatives	SiO₂NPs (5 mg), rt, EtOH	86–98	12
		RCHO + R^1—R^2 + 2 $\begin{smallmatrix}CN\\CN\end{smallmatrix}$ R = aryl R^1 = Me, Ph R^2 = H, Me	Polysubstituted anilines	SiO₂NPs (10 mg), reflux, EtOH	52–65	5
4.	Cu/SB-Fe₃O₄	aldehyde + H_3C-acetoacetate + H_2N-NH_2 urea R = OEt, OMe R^1 = H, Cl, NO₂, OMe, Me, Br	3,4-dihydro pyrimidinone derivatives	Cu/SB-Fe₃O₄ (0.01 g), 60°C, solvent-free	87–96	10

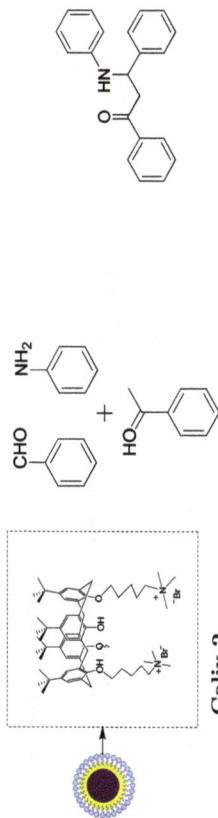

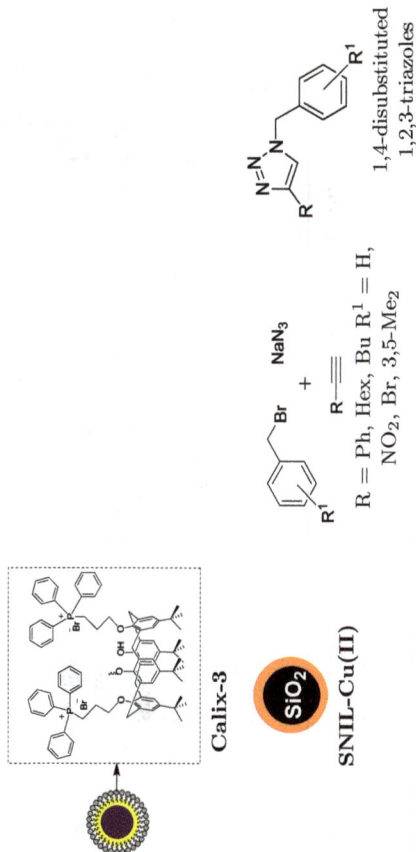

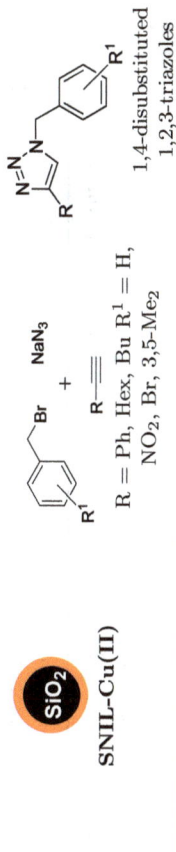

No.	Catalyst	Reaction	Conditions	Yield	Ref
5.	Calix-2 Calix-3	benzaldehyde (CHO) + aniline (NH_2) + acetophenone → 3-oxo-1,3-diphenyl-amine product	Calix-2 (2 mol%), rt, water	69	1
			Calix-3 (2 mol%), rt, water	77	1
6.	SNIL-Cu(II) (SiO_2)	benzyl bromide (R^1) + NaN_3 + R—alkyne R = Ph, Hex, Bu R^1 = H, NO_2, Br, 3,5-Me$_2$ → 1,4-disubstituted 1,2,3-triazoles	SNIL-Cu(II) (0.05 mol%), Na-ascorbate (2 mol%), rt, PEG-400/H_2O	91–99	9

(Continued)

Table 6.2. (*Continued*)

S.No.	Catalyst	Reactants	Product	Conditions	Yield (%)	No. of examples
7.	NS-SSA	NH_2 ... R^1 + H_3C ... OR^2 + CHO ... R; R = H, CH_3, Br, F R^1 = H, Cl, Br, OCH_3, CH_3 R^2 = CH_3, CH_2CH_3	1,2,3,4-tetrahydropyridines	NS-SSA (0.05 g), 65°C, CH_3CN	75–92	13
8.	Fe_3O_4@SiO_2@Propyl–ANDSA	CHO ... R + amino-triazole NH_2 + dimedone	tetrahydrotetrazolo [1,5-a]quinazolines	Fe_3O_4@SiO_2@ Propyl-ANDSA (0.2 g), 100°C, H_2O/EtOH	85–94	6

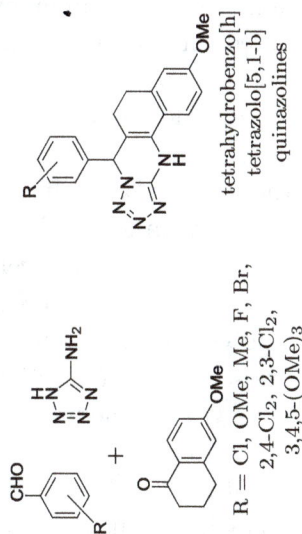

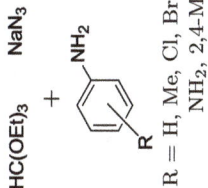

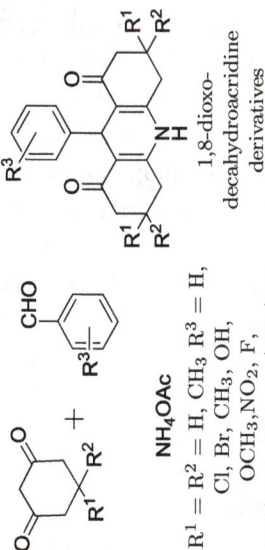

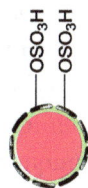

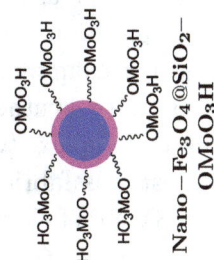

9. Fe₃O₄ @silica sulfonic acid

—OSO₃H
—OSO₃H

10. Nano−Fe₃O₄@SiO₂−OMoO₃H

OMoO₃H   OMoO₃H   OMoO₃H   OMoO₃H   OMoO₃H
HO₃MoO   HO₃MoO   HO₃MoO   OMoO₃H

Row 9

CHO (on benzene ring with R) + triazole-NH₂ + OMe-substituted tetralone

R = Cl, OMe, Me, F, Br, 2,4-Cl₂, 2,3-Cl₂, 3,4,5-(OMe)₃

Product: tetrahydrobenzo[h]tetrazolo[5,1-b]quinazolines

Fe₃O₄@silica sulfonic acid (0.02 g), 100°C, solvent-free

Yield: 85–90   Ref: 8

Row (12)

HC(OEt)₃ + NaN₃ + aniline (R)

R = H, Me, Cl, Br, COCH₃, NH₂, 2,4-Me₂

Product: 1-substituted 1H-tetrazoles

Fe₃O₄@silica sulfonic acid (0.02 g), 100°C, solvent-free

Yield: 78–97   Ref: 12

Row (19)

R^3–CHO + cyclohexanedione (R^1, R^2) + NH₄OAc

$R^1 = R^2$ = H, CH₃ R^3 = H, Cl, Br, CH₃, OH, OCH₃, NO₂, F, 2-Naphthaldehyde, (CH₃)₂N

Product: 1,8-dioxo-decahydroacridine derivatives

Nano-Fe₃O₄@SiO₂–OMoO₃H (0.02 g), 100°C, solvent-free

Yield: 86–94   Ref: 19

A report by Moghaddam *et al.* described the immobilization of glucose onto (3-aminopropyl)triethoxysilane-functionalized SMNPs ([MNPs@GLU][Cl]). Copper salt was further stabilized onto this nanomaterial to make it a potential candidate for the synthesis of different derivatives of 1,2,3-triazole in a one-pot three-component reaction of alkynes, alkyl halides and sodium azide in water (Table 6.2, Entry 2). One of the most attractive features of the catalyst was the use of glucose as the ligand, which renders high dispersibility in water *via* creating hydrogen bondings [59].

4*H*-pyran derivatives are very important from biological and pharmaceutical point of views due to their extensive range of properties including spamolytic, anticancer, anticoagulant and many more. Moreover, these are also administrated in various drugs that are involved in the treatment of neurodegenerative disorders, including Parkinson's disease, Alzheimer's disease and Huntington's disease [60]. Thus, numerous reports documented the synthesis of these compounds, but are associated with the use of toxic solvents and/or amine-based catalysts, elongated reaction time and tedious catalyst synthesis/recovery [61]. In this regard, Sereda *et al.* developed a green and environmentally benign one-pot multicomponent protocol for the synthesis of 4*H*-pyrans and polysubstituted aniline derivatives using SNPs as a catalyst [62]. Three-component reaction of aldehyde, malononitrile and 5,5-dimethyl-1,3-cyclohexanedione or ethyl acetoacetate at room temperature or in refluxing ethanol formed 4*H*-pyran derivatives, while polysubstituted anilines were achieved from four-component reaction of aldehyde, ketone and two equivalents of malononitrile in ethanol (Table 6.2, Entry 3). The salient features of using SiO_2 NPs include simple and inexpensive catalyst preparation and very mild and neutral reaction conditions.

In another example, biologically active 3,4-dihydropyrimidinone derivatives were synthesized using iron oxide-supported copper/ Schiff base complex ($Cu/SB-Fe_3O_4$) (Table 6.2, Entry 4). The nanocatalyst was fabricated by chemical attachment of Schiff base groups on SMNPs followed by treatment with copper salt under mild conditions. The catalyst was easily recovered and reused for 11 cycles with unaltered efficiency and stability [63].

Sayin and Yilmaz [64] synthesized two flexible and bulky groups containing calix[4]arene-based Lewis acid-type catalysts by grafting onto epoxyl-functionalized silica-coated Fe_3O_4 MNPs. The catalytic activity of these catalysts was investigated in a three-component Mannich reaction of benzaldehyde with aniline and acetophenone in the presence of water as solvent (Table 6.2, Entry 5). Out of flexible (calix 2) and bulky (calix 3) groups-substituted calix[4]arene catalyst, the catalytic efficiency of the latter was found to be higher under the same reaction conditions [64].

Recently, Moghaddam *et al.* have developed a stable and active solid-phase catalyst, SNIL-Cu(II) (SNPs-supported copper-containing ionic liquid (IL)), for click reaction between a variety of alkynes, organic halides and sodium azide at room temperature to form 1,4-disubstituted 1,2,3-triazoles (Table 6.2, Entry 6) with high turnover frequency (up to 7920 h^{-1}) [65]. The catalyst was easily synthesized from 1,2-bis(4-pyridylthio)ethane immobilized on SNPs modified with 3-chloropropyltrimethoxysilane and $Cu(OTf)_2$. Furthermore, the catalyst was reused up to six consecutive runs with nearly similar catalytic activity.

In recent years, the use of acid-supported catalysts has gathered considerable attention because of their exclusive features, such as high selectivity, easy workup, convenient separation and greener nature due to less waste generation [66]. Considering these points, Daraei *et al.* [67] have developed nanosphere silica sulfuric acid (NS-SSA) as catalyst for the synthesis of 1,2,3,4-tetrahydropyridines by one-pot MCR of aniline, arylaldehydes and β-ketoester in acetonitrile (Table 6.2, Entry 7). The NS-SSA catalyst was synthesized by the reaction of nanosphere silica with chlorosulfonic acid. A wide range of electron-releasing and electron-withdrawing substituents and halogens on anilines and arylaldehydes and different β-ketoester were tolerated well and afforded the desired tetrahydropyridines in good to excellent yields [67].

Similarly, Ghorbani Vaghei *et al.* [68] have developed 7-aminonaphthalene-1,3-disulfonic acid-functionalized magnetic Fe_3O_4 nanoparticles (Fe_3O_4@SiO_2@Propyl-ANDSA) as a green and effective catalyst for the synthesis of quinazoline derivatives (Table 6.2, Entry 8). The new derivatives of tetrahydrotetrazolo[1,5-a]quinazolines and tetrahydrobenzo[h]tetrazolo[5,1-b] quinazolines were derived in one pot through the condensation reaction of aldehydes, 5-aminotetrazole and dimedone or 6-methoxy-3,4-dihyronaphtalen-1(2H)-one at 100°C in a mixture of water and ethanol. The supported acid catalyst releases H^+ ions that serve as electrophilic species which protonate the aldehydes for Knoevenagel condensation with dimedone giving a benzylidene compound. Benzylidene on subsequent Michael addition with 5-aminotetrazole forms the final ring system [68].

Considering the importance of tetrazoles in material science and coordination chemistry, Naeimi and Mohamadabadi fabricated nanomagnetic solid acid (Fe_3O_4@silica sulfonic acid) for the synthesis of 1-substituted 1*H*-tetrazoles using a one-pot three-component reaction between triethyl orthoformate, an amine and sodium azide [69]. They immobilized $-SO_3H$ groups on core–shell composite by simple mixing of SMNPs and chlorosulfonic acid in CH_2Cl_2 (Table 6.2, Entry 9). All the products were formed in good to excellent yields, short reaction times and under solventless reaction conditions.

Molybdic acid-functionalized silica-coated nano-Fe_3O_4 magnetic particles (nano-Fe_3O_4@SiO_2–$OMoO_3H$) have been fabricated and employed in the synthesis of 1,8-dioxo-decahydroacridine derivatives under solvent-free conditions from aromatic aldehydes, an amine and a dimedone (Table 6.2, Entry 10). The acid catalyst protonates the carbonyl group of aldehyde and thus makes it a convenient electrophile which then condenses with dimedone to initiate the reaction. High stability, remarkable catalytic activity, shorter reaction time, good recyclability and reusability are some of the attractions of the reported methodology [70].

6.2.5 CO_2 capturing

Considering the harmful effects of rising CO_2 level in Earth's atmosphere, researchers have been developing strategies by which CO_2 can be converted into various value-added chemicals [71]. On this note, Sharma and co-workers have fabricated an SMNP-based copper nanocatalyst (Cu-ABF@ASMNP) for the fixation of CO_2 through the cycloaddition reaction of CO_2 and epoxide to form cyclic carbonates at atmospheric pressure (Scheme 6.15) [72]. The salient features of the protocol include mild reaction conditions and reusability of the catalyst for at least five consecutive cycles.

Similarly, Fan *et al.* captured CO_2 by synthesizing diphenyl carbonate using phenol and in the presence of magnetic $Fe_3O_4@SiO_2$–ZnX_2 catalyst (Scheme 6.16) [73]. Among various zinc halides which were screened, supported $ZnBr_2$ gave the best results, even better than its homogeneous counterpart.

Conversion of CO_2 into methane is another step toward reducing global warming [74]. In this regard, Aziz *et al.* have fabricated a mesostructured SNP-supported nickel-based catalyst (Ni/MSN) for CO_2 methanation [75]. The mesoporous structure provided a large surface area which promoted high dispersion of metal species on the surface of the support. Also, the pores of Ni/MSN facilitated higher diffusion of CO_2 which increased the overall reaction rate

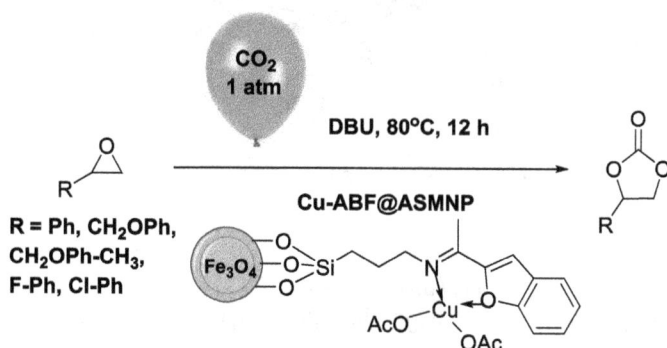

Scheme 6.15. Cycloaddition reaction of CO_2 and epoxide.

Scheme 6.16. Synthesis of diphenyl carbonate using Fe_3O_4@SiO_2–$ZnBr_2$ catalyst.

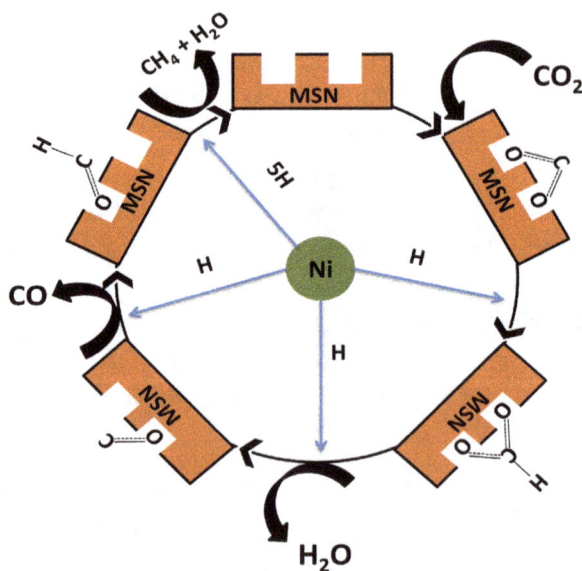

Scheme 6.17. Mechanism for CO_2 methanation.

to 19.16×10^2 mol CH_4/mol Ni s at 573 K, $H_2/CO_2 = 4/1$ and with a selectivity (CH_4) of 99.9%. The authors described a detailed mechanism for the above reaction (Scheme 6.17). First, Ni species, present on Ni/MSN, dissociate H_2 molecule into atomic hydrogen which reacts with the surface oxygen to form water. This results in the formation of oxygen vacancies which activate additional CO_2 to fill the vacancies and produce CO. Further, successive reduction will occur to form CH_4 as the product.

6.2.6 Hydrogen production

Hydrogen gas is considered as one of the most promising energy carriers due its to high energy capacity and environmental friendliness. So, development of new strategies for its release and storage is becoming vital these days [76]. Recently, Dong *et al.* [77] reported the synthesis of hollow mesoporous silica-coated magnetic nanocapsules and further utilized it for hydrogen generation. The Pt nanoparticles were immobilized in the core and were covered with a mesoporous silica shell which prevents the possibility of leaching. The nanocapsules showed superior catalytic activity in the hydrogen generation from hydrolysis of ammonia borane. The cavities in the mesoporous silica shell allowed the increased exposure of active sites and free movement of the molecules, resulting in enhanced catalytic activity [77].

Similarly, Akbayrak *et al.* have developed ruthenium(0) nanoparticle-supported silica-coated cobalt ferrite nanoparticles $(Ru(0)/SiO_2-CoFe_2O_4)$ for hydrogen generation by catalytic hydrolysis of ammonia borane (AB) [78]. For this, first ruthenium(III) ions were impregnated on the surface of $SiO_2-CoFe_2O_4$. Next, Ru(III) ions were reduced to Ru(0) nanoparticles *in situ* from the reduction of $Ru^{3+}/SiO_2-CoFe_2O_4$ during the catalytic hydrolysis of AB. It was observed that when AB solution is added to the suspension of $Ru^{3+}/SiO_2-CoFe_2O_4$, both reduction of ruthenium(III) to ruthenium(0) and hydrogen release from the hydrolysis of AB occur concomitantly. Similar work has also been done using Pd(0) nanoparticles [79].

6.2.7 C–H activation

The direct activation of C–H bond is a highly attractive strategy in covalent synthesis as it circumvents the use of activating/directing groups and minimizes time as well as materials [80]. A magnetically retrievable silica-based copper-catalyzed synthesis of carbamates has been developed *via* C–H activation of formamides

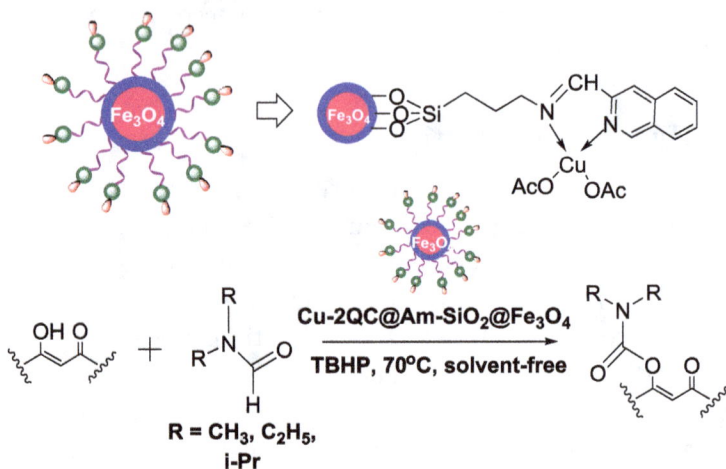

Scheme 6.18. Cu–2QC@Am–SiO$_2$@Fe$_3$O$_4$-catalyzed synthesis of carbamates *via* C–H activation of formamides.

(Scheme 6.18) [81]. The catalyst was fabricated by immobilizing quinoline-2-carboxaldehyde on functionalized SMNPs and further incorporating copper acetate (Cu–2QC@Am–SiO$_2$@Fe$_3$O$_4$). The products so formed have significant industrial and pharmaceutical importance. The method offers numerous advantages, such as mild reaction conditions, broad substrate scope and simple workup procedure.

Recently, considerable amount of interest has grown for nickel-containing complexes with higher oxidation states (III, IV), especially for catalyzing oxidative processes [82]. Till now, only a few examples have been reported that are stable, isolated and completely characterized. This is because complexes having Ni(III, IV) are relatively unstable and light- and moisture-sensitive [83]. Recently, Ni(III) derivative-catalyzed electrochemical oxidation reactions displayed good results under homogeneous conditions [84]. But, in order to obtain a stable Ni complex with higher oxidation state, ligand selection is a critical step, and for this, fluoro-, oxo- and certain macrocyclic N-donor ligands are generally preferred. Another approach to achieve a stable nickel complex is to provide

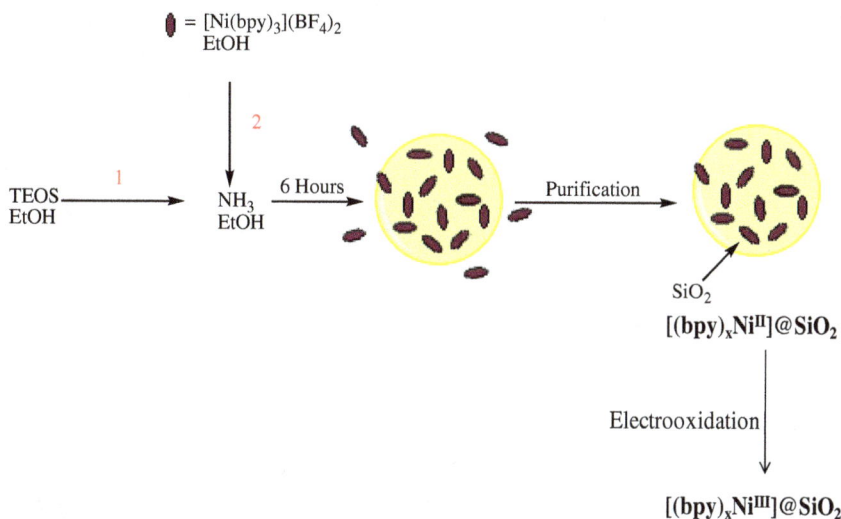

Scheme 6.19. Synthesis of $[(bpy)_xNi^{III}]@SiO_2$.

a special environment that makes the complex effective for performing catalytic reactions, easy handling, convenient isolation and regeneration [85]. Khrizanforov *et al.* [86] developed a green and atom economical technique to fabricate a stable nano-sized nickel (III) complex ($[(bpy)_xNi^{III}]@SiO_2$) with 2,2' bipyridine ligands (bpy) stabilized electrochemically using silicate shell (Scheme 6.19). The synthesized nanocatalyst was utilized in fluoroalkylation reaction of the C–H bonds for various aromatic compounds. It was observed that doping of metal complexes into SNPs generated easily reusable heterogenized nanocatalyst which resulted in 100% conversion of reaction between arenes with perfluoroheptanoic acid in one step [86].

6.2.8   Condensation reaction

Knoevenagel condensation is defined as a reaction between an aldehyde/ketone with any compound having an active methylene group. Ying *et al.* performed the condensation reaction under ultrasonic irradiation [87]. The above strategy has several advantages such as short reaction time which induces high local temperature and pressure, clean reaction condition and high yield and selectivity of the products so formed. To catalyze Knoevenagel condensation of aromatic aldehydes and α-aromatic-substituted methylene compounds using ultrasonication in water, they developed a heterogeneous amine using SMNPs, SiO_2@MNP-A (Scheme 6.20).

Synthesis and immobilization of nanoparticles in ILs have been extensively investigated as ILs are valued key components for catalysts due to their cation–anion interactive nature under ambient temperature conditions [88]. Safaei-Ghomi and co-workers [89] fabricated a nanocatalyst by tethering bis(1(3-trimethoxysilylpropyl)-3-methyl-imidazolium) copper tetrachloride to colloidal SNPs. The synthesized nanocatalyst was employed for preparing trans-2,3-dihydrofurans *via* one-pot condensation reaction of 4-bromophenacyl bromide, aromatic aldehydes and 5,5-dimethyl-1,3-cyclohexanedione under microwave irradiation (Scheme 6.21). The major assets of this work include high efficiency, reusability of catalyst, one-pot diastereoselective synthesis and greener and cleaner microwave procedure [89].

Ying *et al.* [90] have synthesized imidazolium ionic moiety-modified chiral aminocyclohexane in combination with sulfamide supported on MNPs. The catalyst showed great performance in the enantioselective Michael addition and in asymmetric aldol condensation (Scheme 6.22) with excellent yields and stereoselectivities [90]. The IL links the reactants and catalytic sites in water,

Scheme 6.20. Knoevenagel condensation of aromatic aldehydes and α-aromatic-substituted methylene compounds using SiO_2@MNP-A.

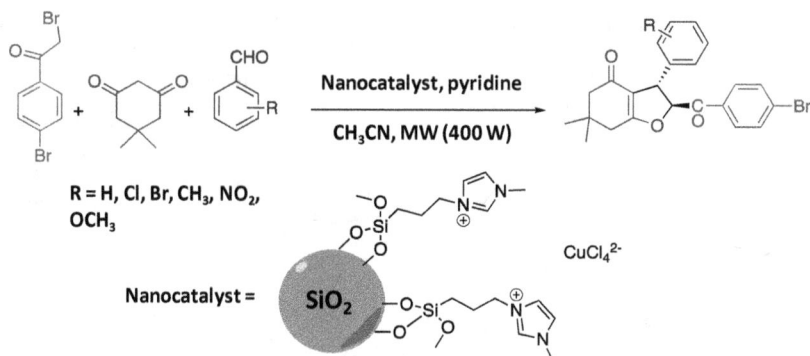

Scheme 6.21. Fabrication of dihydrofurans.

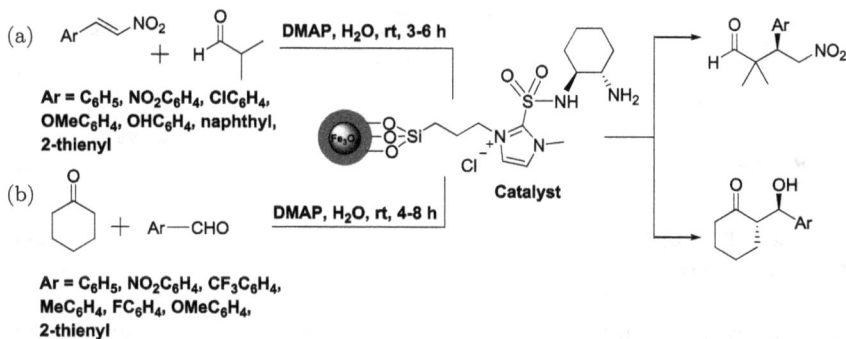

Scheme 6.22. (a) Enantioselective Michael addition and (b) asymmetric aldol condensation reactions.

which makes the reported procedure green and environmentally benign.

Environmentally benign zirconium-based nanocatalyst was developed by Sharma and co-workers [91] *via* covalently grafting 3-hydroxy-2-methyl-1,4-naphthoquinone over the amine-function-alized SMNPs and subsequent complexation with ZrOCl$_2$. The synthesized catalyst (Zr(IV)-HMNQ@ASMPs) showed superior activity for a variety of organic transformations, including Friedel–Crafts, Knoevenagel and Pechmann condensation reactions (Scheme 6.23). High turnover numbers and chemoselectivity, mild reaction conditions,

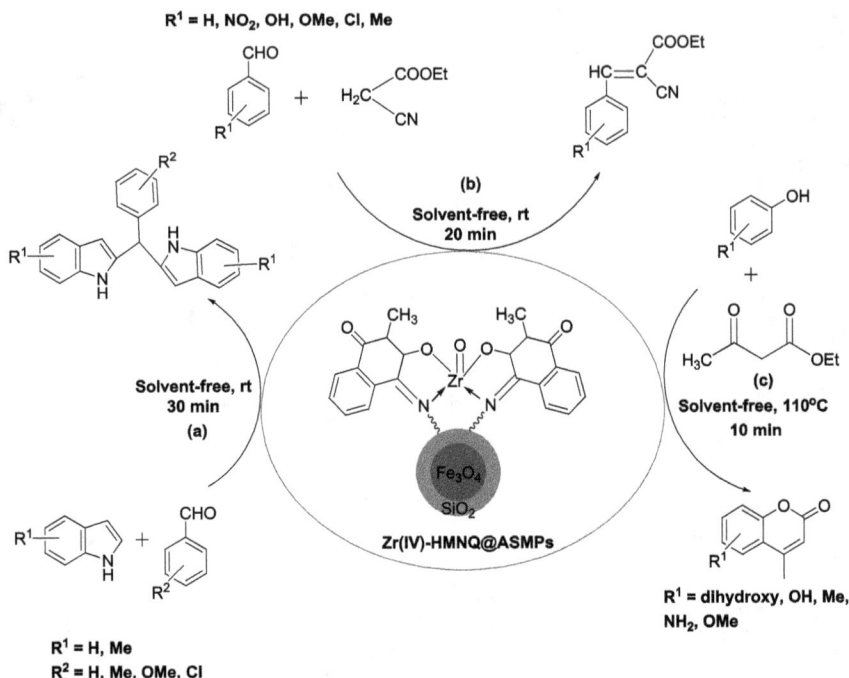

Scheme 6.23. (a) Friedel–Crafts, (b) Knoevenagel and (c) Pechmann condensation reactions catalyzed by Zr(IV)–HMNQ@ASMPs.

simple preparation and economic viability along with excellent recoverability and reusability are some of the advantages of the reported protocol [91].

6.3 Critical Comparison with Other Supported Catalysts

This chapter described the design, fabrication and utilization of SNPS and SMNPs as catalytic supports for a variety of organic transformations. Table 6.3 compares the catalytic features of these nanocomposites with other heterogeneous and homogeneous catalytic systems. The table comprehends that SNP- and SMNP-based catalytic entities not only impart excellent recoverability and reusability but also provide reasonably good activity in comparison to their homogeneous counterparts.

Table 6.3. A critical comparison of SNP- and SMNP-based catalysts with other reported homogeneous and heterogeneous catalysts.

S. No.	Reaction	Catalyst	Catalyst amount	Yield of desired product	Catalyst separation	No. of recovery cycles	Ref.
1	Coupling reaction						
1.1	C–C (Suzuki coupling)	Pd nanocatalyst	0.2 mol%	82–95 yield%	Magnetic separation	16	[16]
		SNP-supported NHC–Pd complex	2 mol%	76–98 yield%	Centrifugation at 750 rpm	5	[92]
		Polymer-supported Pd–NHC complex	1 mol%	91–97 yield%	Filtration	10	[93]
		Pd(OAc)$_2$	0.4 mol%	45–96 yield%	—	—	[94]
1.2	C–N (Amination of aryl halides)	Fe$_3$O$_4$@SiO$_2$Cu	25 mg	74–92 yield%	Magnetic separation	3	[30]
		Polymer-supported Cu(II) catalyst	0.05 g; 0.026 mmol	37–98 yield%	Filtration	5	[95]
		CuI	10 mol%	61–87 yield%	—	—	[96]
1.3	C–O (*O*-arylation of phenols)	Fe$_3$O$_4$@SiO$_2$@PPh$_2$@ Pd(0)	0.005 g; 0.015 mmol of Pd	70–93 yield%	Magnetic separation	6	[32]
		CuI on hybrid silica	5 mol%	65–99 yield%	Filtration	6	[97]

(*Continued*)

Table 6.3. (*Continued*)

S. No.	Reaction	Catalyst	Catalyst amount	Yield of desired product	Catalyst separation	No. of recovery cycles	Ref.
		Cu NPs on microporous covalent triazine polymer	100 mg; 2.6 wt.% Cu	71–97 yield%	Centrifugation	5	[98]
		CuI	10 mol%	60–92 yield%	—	—	[99]
1.4	C–S	Fe$_3$O$_4$@SiO$_2$@C22–Pd(II)	0.5 mol%	13–94 yield%	Magnetic separation	5	[33]
		Cu-grafted functionalized mesoporous SBA-15	25 mg	70–79 yield%	Filtration	6	[100]
		Alumina-supported copper sulfate	5 mol%	70–98 yield%	Centrifugation/filtration	8	[101]
		CuI	5 mol%	33–98 yield%	—	—	[102]
2	Oxidation of alcohols	SiO$_2$@Fe$_3$O$_4$-Pro-Pd	0.5 mol% Pd	58–97 yield%	Magnetic separation	8	[103]
		Palladium/silica	3 mol% of Pd	40–96 yield%	Filtration	5	[104]
		Hydroxyapatite-supported Pd nanoclusters	0.1 g; Pd 0.2 mol%	80–99 yield%	Centrifugation/filtration	3	[105]
		Pd(OAc)$_2$	0.05 mmol	71–100 yield%	—	—	[106]
3	Reduction of nitroarenes	Cu–AcTp@Am–Si–Fe$_3$O$_4$	20 mg	90–100 selectivity%	Magnetic separation	9	[55]
		Au/SiO$_2$	0.1 g	65.4–100 selectivity%	—	—	[107]

		AgNPs-MPTA	0.05 g	81–98 yield%	Filtration	5	[108]
		Cu nanoparticles	3 mmol	75–90 yield%	—	—	[109]
4	CO_2 capturing (formation of cyclic carbonate from epoxides and CO_2)	Cu-ABF@ASMNPs	50 mg	89–99 yield%	Magnetic separation	5	[72]
		Silica-supported 4-pyrrolidinopyridinium iodide	3.8×10^{-2} mmol	80–89 yield%	Filtration	3	[110]
		Cross-linked PVP-supported $ZnBr_2$	0.29 g	53.1 yield%	Filtration	7	[111]
		Binaphthyldiamino Cu(II) Salen complex	4.5×10^{-5} mol	89–100 yield%	—	—	[112]
5	Hydrogen production	Ru(0)/SiO_2–$CoFe_2O_4$	10 mg	172 min^{-1} TOF	Magnetic separation	10	[78]
		RuNPs@ZK-4	18.2 mg	5410 h^{-1} TOF	Filtration/centrifugation	5	[113]

(*Continued*)

Table 6.3. (*Continued*)

S. No.	Reaction	Catalyst	Catalyst amount	Yield of desired product	Catalyst separation	No. of recovery cycles	Ref.
6	C–H activation	Cu–2QC@Am–SiO$_2$@Fe$_3$O$_4$	30 mg	60–99 conversion%	Magnetic separation	8	[81]
		Cu(OAc)$_2$	5 mol%	69–86 yield%	—	—	[114]
7	Knoevenagel condensation	Zr(IV)–HMNQ@ASMPs	20 mg	93–99 conversion%	Magnetic separation	5	[91]
		Nitridated fibrous silica KCC-1-N500	30 mg	45–99 conversion%	Filtration/centrifugation	5	[115]
		AlPO$_4$–Al$_2$O$_3$	1.5–3 g	43–89 yield%	—	—	[116]
		ZrOCl$_2$·8H$_2$O	0.1 g	73–98 yield%	—	—	[117]

6.4 Conclusion

Nanotechnology has reformulated the definition of catalysis synthesis. Nanocatalysis constitutes a new and advanced class of materials which has provided numerous economic and environmental benefits. This chapter illustrates the promise and potential of silica-based nanocatalysts in a variety of organic reactions. Besides improving the stability, silica support provides a magnificent platform for surface modification that strongly influences the catalytic activity in addition to easy recovery and recyclability. Hence, SNPs and silica-encapsulated MNPs have displayed remarkable results as supports in various catalytic reactions, such as coupling, oxidation, reduction, multicomponent, condensation and CO_2 capturing. However, sincere efforts are required in some of the other industrially significant and emerging organic transformations including asymmetric synthesis, biomass conversion, photocatalytic reactions, microwave-assisted reactions, biocatalysis and vapor-phase catalysis. Besides, massive attention should be given to some of the missing scientific essentials such as mechanistic studies through *in situ* catalysis and multiplication of products yield at the industrial level. Owing to the high stability and strong interaction between support and metal, silica-based nanomaterials can offer a powerful platform for future developments in catalytic field, as elaborated in Chapter 7.

References

[1] (a) P. T. Anastas, L. B. Bartlett, M. M. Kirchhoff, T. C. Williamson, *Catalysis Today* **2000**, *55*, 11–22; (b) C. A. Busacca, D. R. Fandrick, J. J. Song, C. H. Senanayake, *Advanced Synthesis & Catalysis* **2011**, *353*, 1825–1864.

[2] (a) R. N. Baig, R. S. Varma, *Chemical Communications* **2013**, *49*, 752–770; (b) N. Mizuno, M. Misono, *Chemical Reviews* **1998**, *98*, 199–218.

[3] (a) D. Astruc, F. Lu, J. R. Aranzaes, *Angewandte Chemie International Edition* **2005**, *44*, 7852–7872; (b) A. T. Bell, *Science* **2003**, *299*, 1688–1691.

[4] (a) S. Francis, S. Joseph, E. P. Koshy, B. Mathew, *Environmental Science and Pollution Research* **2017**, *24*, 17347–17357; (b) A. Bazgir, G. Hosseini, R. Ghahremanzadeh, *ACS Combinatorial Science* **2013**, *15*, 530–534; (c) W.-N. Wang, W.-J. An, B. Ramalingam, S. Mukherjee, D. M. Niedzwiedzki, S. Gangopadhyay, P. Biswas, *Journal of the American*

Chemical Society **2012**, *134*, 11276–11281; (d) K. Kwak, W. Choi, Q. Tang, M. Kim, Y. Lee, D.-e. Jiang, D. Lee, *Nature Communications* **2017**, *8*, 14723; (e) B. Gurunathan, A. Ravi, *Bioresource Technology* **2015**, *188*, 124–127.

[5] (a) M. B. Gawande, R. Zboril, V. Malgras, Y. Yamauchi, *Journal of Materials Chemistry A* **2015**, *3*, 8241–8245; (b) S. C. Tsang, V. Caps, I. Paraskevas, D. Chadwick, D. Thompsett, *Angewandte Chemie* **2004**, *116*, 5763–5767; (c) Q. M. Kainz, O. Reiser, *Accounts of Chemical Research* **2014**, *47*, 667–677.

[6] (a) Y. Piao, A. Burns, J. Kim, U. Wiesner, T. Hyeon, *Advanced Functional Materials* **2008**, *18*, 3745–3758; (b) C. Gonzalez-Arellano, A. M. Balu, R. Luque, D. J. Macquarrie, *Green Chemistry* **2010**, *12*, 1995–2002.

[7] A. Bitar, N. M. Ahmad, H. Fessi, A. Elaissari, *Drug Discovery Today* **2012**, *17*, 1147–1154.

[8] I. I. Slowing, B. G. Trewyn, S. Giri, V. Y. Lin, *Advanced Functional Materials* **2007**, *17*, 1225–1236.

[9] P. Couleaud, V. Morosini, C. Frochot, S. Richeter, L. Raehm, J.-O. Durand, *Nanoscale* **2010**, *2*, 1083–1095.

[10] (a) S. Giri, B. G. Trewyn, M. P. Stellmaker, V. S. Y. Lin, *Angewandte Chemie International Edition* **2005**, *44*, 5038–5044; (b) J. Kim, H. S. Kim, N. Lee, T. Kim, H. Kim, T. Yu, I. C. Song, W. K. Moon, T. Hyeon, *Angewandte Chemie International Edition* **2008**, *47*, 8438–8441.

[11] (a) M. B. Gawande, Y. Monga, R. Zboril, R. Sharma, *Coordination Chemistry Reviews* **2015**, *288*, 118–143; (b) R. K. Sharma, S. Sharma, S. Dutta, R. Zboril, M. B. Gawande, *Green Chemistry* **2015**, *17*, 3207–3230; (c) R. K. Sharma, M. Yadav, M. B. Gawande, *Ferrites and Ferrates: Chemistry and Applications in Sustainable Energy and Environmental Remediation*, ACS Publications, **2016**, pp. 1–38; (d) R. K. Sharma, S. Dutta, S. Sharma, R. Zboril, R. S. Varma, M. B. Gawande, *Green Chemistry* **2016**, *18*, 3184–3209.

[12] (a) D. E. De Vos, M. Dams, B. F. Sels, P. A. Jacobs, *Chemical Reviews* **2002**, *102*, 3615–3640; (b) A. Wight, M. Davis, *Chemical Reviews* **2002**, *102*, 3589–3614.

[13] S.-H. Wu, C.-Y. Mou, H.-P. Lin, *Chemical Society Reviews* **2013**, *42*, 3862–3875.

[14] S. Shylesh, A. Wagner, A. Seifert, S. Ernst, W. R. Thiel, *Chemistry — A European Journal* **2009**, *15*, 7052–7062.

[15] (a) N. Miyaura, A. Suzuki, *Chemical Reviews* **1995**, *95*, 2457–2483; (b) R. Martin, S. L. Buchwald, *Accounts of Chemical Research* **2008**, *41*, 1461–1473.

[16] T. Azadbakht, M. A. Zolfigol, R. Azadbakht, V. Khakyzadeh, D. M. Perrin, *New Journal of Chemistry* **2015**, *39*, 439–444.

[17] A. Dadras, M. R. Naimi-Jamal, F. M. Moghaddam, S. E. Ayati, *Applied Organometallic Chemistry* **2018**, *32*, e3993.

[18] A. Khazaei, M. Khazaei, M. Nasrollahzadeh, *Tetrahedron* **2017**, *73*, 5624–5633.

[19] R. K. Sharma, M. Yadav, R. Gaur, Y. Monga, A. Adholeya, *Catalysis Science & Technology* **2015**, *5*, 2728–2740.

[20] A. Hajipour, G. Azizi, *Applied Organometallic Chemistry* **2015**, *29*, 247–253.

[21] W. Yang, L. Wei, F. Yi, M. Cai, *Tetrahedron* **2016**, *72*, 4059–4067.

[22] S. Thorand, N. Krause, *The Journal of Organic Chemistry* **1998**, *63*, 8551–8553.

[23] R. K. Sharma, M. Yadav, R. Gaur, R. Gupta, A. Adholeya, M. B. Gawande, *ChemPlusChem* **2016**, *81*, 1312–1319.

[24] S. Banerjee, J. Das, S. Santra, *Tetrahedron Letters* **2009**, *50*, 124–127.

[25] S. Banerjee, S. Santra, *Tetrahedron Letters* **2009**, *50*, 2037–2040.

[26] L. Jiang, S. L. Buchwald, *Metal-Catalyzed Cross-Coupling Reactions*, 2nd edn., Wiley-VCH, Weinheim, **2004**, pp. 699–760.

[27] S. M. Inamdar, V. K. More, S. K. Mandal, *Chemistry Letters* **2012**, *41*, 1484–1486.

[28] L. Aurelio, R. T. Brownlee, A. B. Hughes, *Chemical Reviews* **2004**, *104*, 5823–5846.

[29] R. K. Sharma, Y. Monga, A. Puri, G. Gaba, *Green Chemistry* **2013**, *15*, 2800–2809.

[30] R. N. Baig, R. S. Varma, *RSC Advances* **2014**, *4*, 6568–6572.

[31] E. Buck, Z. J. Song, D. Tschaen, P. G. Dormer, R. Volante, P. J. Reider, *Organic Letters* **2002**, *4*, 1623–1626.

[32] M. A. Zolfigol, V. Khakyzadeh, A. R. Moosavi-Zare, A. Rostami, A. Zare, N. Iranpoor, M. H. Beyzavi, R. Luque, *Green Chemistry* **2013**, *15*, 2132–2140.

[33] B. Movassagh, A. Takallou, A. Mobaraki, *Journal of Molecular Catalysis A: Chemical* **2015**, *401*, 55–65.

[34] (a) R. A. Sheldon, I. Arends, A. Dijksman, *Catalysis Today* **2000**, *57*, 157–166; (b) R. Sheldon, *Metal-Catalyzed Oxidations of Organic Compounds: Mechanistic Principles and Synthetic Methodology Including Biochemical Processes*, Elsevier, **2012**.

[35] M. Hudlicky, *Oxidations in Organic Chemistry*, American Chemical Society, **1990**.

[36] L. Chen, B. Li, D. Liu, *Catalysis Letters* **2014**, *144*, 1053–1061.

[37] (a) B. Meunier, *Chemical Reviews* **1992**, *92*, 1411–1456; (b) K. M. Kadish, K. M. Smith, R. Guilard, *The Porphyrin Handbook: Inorganic, Organometallic and Coordination Chemistry*, Vol. 3, Elsevier, **2000**.

[38] (a) E. Brulé, Y. R. de Miguel, *Organic & Biomolecular Chemistry* **2006**, *4*, 599–609; (b) V. R. Rani, M. R. Kishan, S. Kulkarni, K. Raghavan, *Catalysis Communications* **2005**, *6*, 531–538.

[39] A. Rezaeifard, P. Farshid, M. Jafarpour, G. K. Moghaddam, *RSC Advances* **2014**, *4*, 9189–9196.

[40] R. K. Sharma, M. Yadav, Y. Monga, R. Gaur, A. Adholeya, R. Zboril, R. S. Varma, M. B. Gawande, *ACS Sustainable Chemistry & Engineering* **2016**, *4*, 1123–1130.

[41] R. K. Sharma, Y. Monga, *Applied Catalysis A: General* **2013**, *454*, 1–10.

[42] D. Munz, D. Wang, M. M. Moyer, M. S. Webster-Gardiner, P. Kunal, D. Watts, B. G. Trewyn, A. N. Vedernikov, T. B. Gunnoe, *ACS Catalysis* **2016**, *6*, 4584–4593.

[43] Z. Shi, C. Zhang, C. Tang, N. Jiao, *Chemical Society Reviews* **2012**, *41*, 3381–3430.

[44] Y. Fang, Y. Chen, X. Li, X. Zhou, J. Li, W. Tang, J. Huang, J. Jin, J. Ma, *Journal of Molecular Catalysis A: Chemical* **2014**, *392*, 16–21.

[45] J. Sun, G. Yu, L. Liu, Z. Li, Q. Kan, Q. Huo, J. Guan, *Catalysis Science & Technology* **2014**, *4*, 1246–1252.

[46] R. Gaur, M. Yadav, R. Gupta, G. Arora, P. Rana, R. K. Sharma, *ChemistrySelect* **2018**, *3*, 2502–2508.

[47] A. Bayat, M. Shakourian-Fard, N. Ehyaei, M. M. Hashemi, *RSC Advances* **2015**, *5*, 22503–22509.

[48] Z. S. Qureshi, P. B. Sarawade, I. Hussain, H. Zhu, H. Al-Johani, D. H. Anjum, M. N. Hedhili, N. Maity, V. D'Elia, J. M. Basset, *ChemCatChem* **2016**, *8*, 1671–1678.

[49] Y. Cao, W. Zhai, X. Zhang, S. Li, L. Feng, Y. Wei, *ISRN Nanomaterials* **2013**, *2013*, 745397.

[50] (a) F. Fehsenfeld, J. Calvert, R. Fall, P. Goldan, A. B. Guenther, C. N. Hewitt, B. Lamb, S. Liu, M. Trainer, H. Westberg, *Global Biogeochemical Cycles* **1992**, *6*, 389–430; (b) M. Phillips, K. Gleeson, J. M. B. Hughes, J. Greenberg, R. N. Cataneo, L. Baker, W. P. McVay, *The Lancet* **1999**, *353*, 1930–1933.

[51] L. Liotta, *Applied Catalysis B: Environmental* **2010**, *100*, 403–412.

[52] M. Popova, A. Ristić, K. Lazar, D. Maučec, M. Vassileva, N. Novak Tušar, *ChemCatChem* **2013**, *5*, 986–993.

[53] S. A. Lawrence, *Amines: Synthesis, Properties and Applications*, Cambridge University Press, **2004**.

[54] (a) X. Cui, F. Shi, Y. Deng, *ChemCatChem* **2012**, *4*, 333–336; (b) S. K. Ghosh, M. Mandal, S. Kundu, S. Nath, T. Pal, *Applied Catalysis A: General* **2004**, *268*, 61–66.

[55] R. K. Sharma, Y. Monga, A. Puri, *Journal of Molecular Catalysis A: Chemical* **2014**, *393*, 84–95.

[56] R. K. Sharma, S. Dutta, S. Sharma, *New Journal of Chemistry* **2016**, *40*, 2089–2101.

[57] (a) H. Bienaymé, C. Hulme, G. Oddon, P. Schmitt, *Chemistry — A European Journal* **2000**, *6*, 3321–3329; (b) E. Ruijter, R. Scheffelaar, R. V. Orru, *Angewandte Chemie International Edition* **2011**, *50*, 6234–6246.

[58] A. Khojastehnezhad, M. Rahimizadeh, H. Eshghi, F. Moeinpour, M. Bakavoli, *Chinese Journal of Catalysis* **2014**, *35*, 376–382.

[59] F. M. Moghaddam, V. Saberi, S. Kalhor, S. E. Ayati, *RSC Advances* **2016**, *6*, 80234–80243.

[60] (a) L. Bonsignore, G. Loy, D. Secci, A. Calignano, *European Journal of Medicinal Chemistry* **1993**, *28*, 517–520; (b) T.-S. Jin, A.-Q. Wang, X. Wang, J.-S. Zhang, T.-S. Li, *Synlett* **2004**, *2004*, 0871–0873.

[61] (a) A. T. Khan, M. Lal, S. Ali, M. M. Khan, *Tetrahedron Letters* **2011**, *52*, 5327–5332; (b) N. Azizi, S. Dezfooli, M. Khajeh, M. M. Hashemi, *Journal of Molecular Liquids* **2013**, *186*, 76–80; (c) X.-S. Wang, D.-Q. Shi, S.-J. Tu, C.-S. Yao, *Synthetic Communications* **2003**, *33*, 119–126.

[62] S. Banerjee, A. Horn, H. Khatri, G. Sereda, *Tetrahedron Letters* **2011**, *52*, 1878–1881.

[63] D. Elhamifar, P. Mofatehnia, M. Faal, *Journal of Colloid and Interface Science* **2017**, *504*, 268–275.

[64] S. Sayin, M. Yilmaz, *RSC Advances* **2017**, *7*, 10748–10756.

[65] M. Tavassoli, A. Landarani-Isfahani, M. Moghaddam, S. Tangestaninejad, V. Mirkhani, I. Mohammadpoor-Baltork, *ACS Sustainable Chemistry & Engineering* **2016**, *4*, 1454–1462.

[66] (a) Y. Izumi, N. Natsume, H. Takamine, I. Tamaoki, K. Urabe, *Bulletin of the Chemical Society of Japan* **1989**, *62*, 2159–2162; (b) S. Iimura, D. Nobutou, K. Manabe, S. Kobayashi, *Chemical Communications* **2003**, 1644–1645.

[67] M. Daraei, M. A. Zolfigol, F. Derakhshan-Panah, M. Shiri, H. G. Kruger, M. Mokhlesi, *Journal of the Iranian Chemical Society* **2015**, *12*, 855–861.

[68] R. Ghorbani-Vaghei, S. Alavinia, N. Sarmast, *Applied Organometallic Chemistry* **2018**, *32*, e4038.

[69] H. Naeimi, S. Mohamadabadi, *Dalton Transactions* **2014**, *43*, 12967–12973.

[70] M. Kiani, M. Mohammadipour, *RSC Advances* **2017**, *7*, 997–1007.

[71] (a) M. Mikkelsen, M. Jørgensen, F. C. Krebs, *Energy & Environmental Science* **2010**, *3*, 43–81; (b) M. Cokoja, C. Bruckmeier, B. Rieger, W. A. Herrmann, F. E. Kühn, *Angewandte Chemie International Edition* **2011**, *50*, 8510–8537.

[72] R. K. Sharma, R. Gaur, M. Yadav, A. Goswami, R. Zbořil, M. B. Gawande, *Scientific Reports* **2018**, *8*, 1901.

[73] G. Fan, S. Luo, Q. Wu, T. Fang, J. Li, G. Song, *RSC Advances* **2015**, *5*, 56478–56485.

[74] K. R. Thampi, J. Kiwi, M. Graetzel, *Nature* **1987**, *327*, 506.

[75] M. Aziz, A. Jalil, S. Triwahyono, R. Mukti, Y. Taufiq-Yap, M. Sazegar, *Applied Catalysis B: Environmental* **2014**, *147*, 359–368.

[76] (a) A. Züttel, A. Borgschulte, L. Schlapbach, *Hydrogen as a Future Energy Carrier*, John Wiley & Sons, **2011**; (b) K. Mazloomi, C. Gomes, *Renewable and Sustainable Energy Reviews* **2012**, *16*, 3024–3033.

[77] D. Xu, Z. Cui, J. Yang, M. Yuan, X. Cui, X. Zhang, Z. Dong, *International Journal of Hydrogen Energy* **2017**, *42*, 27034–27042.

[78] S. Akbayrak, M. Kaya, M. Volkan, S. Özkar, *Journal of Molecular Catalysis A: Chemical* **2014**, *394*, 253–261.

[79] S. Akbayrak, M. Kaya, M. Volkan, S. Özkar, *Applied Catalysis B: Environmental* **2014**, *147*, 387–393.

[80] (a) A. McNally, B. Haffemayer, B. S. Collins, M. J. Gaunt, *Nature* **2014**, *510*, 129; (b) C. S. Yeung, V. M. Dong, *Chemical Reviews* **2011**, *111*, 1215–1292; (c) D. Kalyani, N. R. Deprez, L. V. Desai, M. S. Sanford, *Journal of the American Chemical Society* **2005**, *127*, 7330–7331.

[81] R. K. Sharma, S. Dutta, S. Sharma, *Dalton Transactions* **2015**, *44*, 1303–1316.

[82] (a) S. Z. Tasker, E. A. Standley, T. F. Jamison, *Nature* **2014**, *509*, 299; (b) M. I. Lipschutz, T. D. Tilley, *Angewandte Chemie International Edition* **2014**, *53*, 7290–7294; (c) N. M. Camasso, M. S. Sanford, *Science* **2015**, *347*, 1218–1220.

[83] Y. B. Dudkina, D. Y. Mikhaylov, T. V. Gryaznova, O. G. Sinyashin, D. A. Vicic, Y. H. Budnikova, *European Journal of Organic Chemistry* **2012**, *2012*, 2114–2117.

[84] M. Khrizanforov, T. Gryaznova, D. Y. Mikhailov, Y. H. Budnikova, O. Sinyashin, *Russian Chemical Bulletin* **2012**, *61*, 1560–1563.

[85] (a) R. Mitra, K. R. Pörschke, *Angewandte Chemie International Edition* **2015**, *54*, 7488–7490; (b) V. M. Iluc, G. L. Hillhouse, *Journal of the American Chemical Society* **2014**, *136*, 6479–6488.

[86] M. N. Khrizanforov, S. V. Fedorenko, S. O. Strekalova, K. V. Kholin, A. R. Mustafina, M. Y. Zhilkin, V. V. Khrizanforova, Y. N. Osin, V. V. Salnikov, T. V. Gryaznova, *Dalton Transactions* **2016**, *45*, 11976–11982.

[87] A. Ying, L. Wang, F. Qiu, H. Hu, J. Yang, *Comptes Rendus Chimie* **2015**, *18*, 223–232.

[88] (a) C. P. Mehnert, R. A. Cook, N. C. Dispenziere, M. Afeworki, *Journal of the American Chemical Society* **2002**, *124*, 12932–12933; (b) A. Riisagera, R. Fehrmanna, M. Haumannb, P. Wasscheidb, *Topics in Catalysis* **2006**, *40*, 91–102.

[89] J. Safaei-Ghomi, H. Shahbazi-Alavi, S. H. Nazemzadeh, *Journal of Nanostructure in Chemistry* **2017**, *7*, 113–119.

[90] A. Ying, S. Liu, Z. Li, G. Chen, J. Yang, H. Yan, S. Xu, *Advanced Synthesis & Catalysis* **2016**, *358*, 2116–2125.

[91] R. K. Sharma, Y. Monga, A. Puri, *Catalysis Communications* **2013**, *35*, 110–114.

[92] S. Tandukar, A. Sen, *Journal of Molecular Catalysis A: Chemical* **2007**, *268*, 112–119.

[93] J.-H. Kim, J.-W. Kim, M. Shokouhimehr, Y.-S. Lee, *The Journal of Organic Chemistry* **2005**, *70*, 6714–6720.

[94] N. E. Leadbeater, M. Marco, *Organic Letters* **2002**, *4*, 2973–2976.

[95] S. M. Islam, S. Mondal, P. Mondal, A. S. Roy, K. Tuhina, N. Salam, M. Mobarak, *Journal of Organometallic Chemistry* **2012**, *696*, 4264–4274.

[96] D. Ma, Q. Cai, H. Zhang, *Organic Letters* **2003**, *5*, 2453–2455.

[97] S. Benyahya, F. Monnier, M. W. C. Man, C. Bied, F. Ouazzani, M. Taillefer, *Green Chemistry* **2009**, *11*, 1121–1123.

[98] P. Puthiaraj, W.-S. Ahn, *Catalysis Science & Technology* **2016**, *6*, 1701–1709.

[99] Y.-H. Liu, G. Li, L.-M. Yang, *Tetrahedron Letters* **2009**, *50*, 343–346.

[100] J. Mondal, P. Borah, A. Modak, Y. Zhao, A. Bhaumik, *Organic Process Research & Development* **2013**, *18*, 257–265.

[101] S. Bhadra, B. Sreedhar, B. C. Ranu, *Advanced Synthesis & Catalysis* **2009**, *351*, 2369–2378.

[102] W. Deng, Y. Zou, Y.-F. Wang, L. Liu, Q.-X. Guo, *Synlett* **2004**, *2004*, 1254–1258.

[103] L. Zhang, P. Li, J. Yang, M. Wang, L. Wang, *ChemPlusChem* **2014**, *79*, 217–222.

[104] D. Choudhary, S. Paul, R. Gupta, J. H. Clark, *Green Chemistry* **2006**, *8*, 479–482.

[105] K. Mori, T. Hara, T. Mizugaki, K. Ebitani, K. Kaneda, *Journal of the American Chemical Society* **2004**, *126*, 10657–10666.

[106] T. Nishimura, T. Onoue, K. Ohe, S. Uemura, *The Journal of Organic Chemistry* **1999**, *64*, 6750–6755.

[107] Y. Chen, J. Qiu, X. Wang, J. Xiu, *Journal of Catalysis* **2006**, *242*, 227–230.

[108] N. Salam, B. Banerjee, A. S. Roy, P. Mondal, S. Roy, A. Bhaumik, S. M. Islam, *Applied Catalysis A: General* **2014**, *477*, 184–194.

[109] A. Saha, B. Ranu, *The Journal of Organic Chemistry* **2008**, *73*, 6867–6870.

[110] K. Motokura, S. Itagaki, Y. Iwasawa, A. Miyaji, T. Baba, *Green Chemistry* **2009**, *11*, 1876–1880.

[111] H. S. Kim, J. J. Kim, H. N. Kwon, M. J. Chung, B. G. Lee, H. G. Jang, *Journal of Catalysis* **2002**, *205*, 226–229.

[112] Y.-M. Shen, W.-L. Duan, M. Shi, *The Journal of Organic Chemistry* **2003**, *68*, 1559–1562.

[113] M. Zahmakiran, *Materials Science and Engineering: B* **2012**, *177*, 606–613.

[114] G. S. Kumar, C. U. Maheswari, R. A. Kumar, M. L. Kantam, K. R. Reddy, *Angewandte Chemie International Edition* **2011**, *50*, 11748–11751.

[115] M. Bouhrara, C. Ranga, A. Fihri, R. R. Shaikh, P. Sarawade, A.-H. Emwas, M. N. Hedhili, V. Polshettiwar, *ACS Sustainable Chemistry & Engineering* **2013**, *1*, 1192–1199.

[116] J. A. Cabello, J. M. Campelo, A. Garcia, D. Luna, J. M. Marinas, *The Journal of Organic Chemistry* **1984**, *49*, 5195–5197.

[117] T. Yakaiah, G. V. Reddy, B. Lingaiah, P. S. Rao, B. Narsaiah, *Indian Journal of Chemistry* **2005**, *44B*, 1301–1303.

Chapter 7

Other Potential Catalytic Applications and Future Perspectives

Gunjan Arora, Sriparna Dutta, Radhika Gupta
and Rakesh Kumar Sharma*
*rksharmagreenchem@hotmail.com

7.1 Introduction

In this new millennium, the field of chemical sciences has embraced burgeoning concepts of "nanotechnology" and specifically "nanocatalysis" to meet the modern challenges of energy and sustainability [1]. Within this context, it is worth mentioning that silica-based organic–inorganic hybrid nanostructured catalytic systems employing either silica nanoparticles (SNPs) or silica-encapsulated magnetite nanoparticles (NPs) have offered undeniable benefits of boosting numerous industrially significant organic transformations by exhibiting superior performance, i.e. maximum selectivity, enormously great activity, low energy consumption and protracted lifetime as already discussed in the previous chapters [2]. What is even worth appreciating is that these precisely engineered third-generation nanocatalysts have been found to be much more effectual and profitable as compared to the conventional catalysts owing to their larger total exposed surface areas and electronic confinement

---

*Corresponding author.

effects within NPs that have raised the possibility of tuning catalytic processes according to the needs. These catalysts have provided a new frontier between homogeneous and heterogeneous catalysis that are the conventional categories of catalysts and have been often regarded as "quasihomogeneous". Undoubtedly, nanotechnology has provided an effective means of controlling the surface structure and properties of supported nanocatalysts without changing their composition. From this discussion, it is quite obvious: *The smaller, the better*. In fact, the effective utilization of the nano-sized catalytic systems has caused a radical revolution in every sector, especially in the fine chemical manufacturing sector (chemical industries) [3]. The global market of nanocatalysis research has revealed that the sales have already reached USD 2,900 which is expected to grow exponentially within few years.

Today, silica-supported nanocatalysts are not only being used for expediting organic transformations but indeed their applicability has seen a wide expansion beginning from biomass conversion to water purification *via* advanced oxidation processes due to their riveting physicochemical properties [4]. The silica-based magnetic nanocomposites in particular have shown promising potential to even deal with environmental challenges because of advantages of low toxicity and biocompatibility in addition to excellent separation properties and the ability to get functionalized in different ways. The role of these nanocomposites is very crucial here as they are helping in the environmental clean-up which also has an intrinsic economic value. This is especially important in view of the fact that global development has reached a stage where the limits of natural resources as well as the environmental impact of human activity present boundaries for future development. In this chapter, we aim to provide a succinct overview of the other potential applications of these versatile nanocatalysts in environmental challenges, particularly in photocatalysis and photodegradation of pollutants in water, biocatalysis, biomass conversion, etc. At the end, we would be discussing about the future challenges confronting the nanocatalysis research that need to be addressed.

7.2 Potential Catalytic Applications of Silica-Based Nanomaterials

7.2.1 Asymmetric catalysis

Asymmetric organocatalysis has witnessed an incredible rise in popularity [5]. The direct employment of pure organic compounds as chiral catalysts has several striking advantages, such as the stability, easy availability, simple handling and storage, which permit most reactions to be performed under non-inert conditions: wet solvent and in air [6]. However, the challenging separation and recycling of organocatalysts makes the reaction non-green and uneconomic. To solve this key issue, heterogenization of homogeneous catalyst onto solid support has gathered tremendous attention. Among various supports, polymer [7], carbon [8] and other metal oxides [9] are well explored for the immobilization of active chiral catalyst. However, these catalysts often suffer from drawbacks, such as complicated preparation of monomeric ligands and time-consuming multistep fabrication protocols for support synthesis [10]. In this scenario, the development of silica-based nanomaterials as solid support would be a smart alternative. Silica-supported core–shell heterogeneous catalysts have relatively large surface area, tunable pore dimensions and well-defined pore arrangement. The core is suitable for accumulating chiral functionalities and the shell is valuable to prevent leaching of chiral organometallics. These characteristics promote easy control over the dispersibility of active species and also facilitate the adjustment of chiral microenvironment of active centers, and thus demonstrate superiority in stereocontrol performance [11].

Considering the above points, Zhang *et al.* have fabricated a rhodium-based core–shell-structured heterogeneous catalyst that effectively catalyzed the asymmetric transfer hydrogenation of aromatic ketones in aqueous medium (Scheme 7.1). The catalyst was synthesized by assembling a chiral Cp*RhTsDPEN complex (Cp* = pentamethyl cyclopentadiene, TsDPEN = 4-methylphenylsulfonyl-1,2-diphenylethylenediamine) within the core of core–shell-structured mesoporous silica spheres (CSSMSS).

Cp*RhTsDPEN-CSSDMSS

Ar = Ph, 4-FPh,4-ClPh, 4-BrPh, 3-BrPh,
4-MePh, 4-OMePh, 3-OMePh, 4-CNPh, 4-CF₃Ph,
4-NO₂Ph, 2-Naphthyl

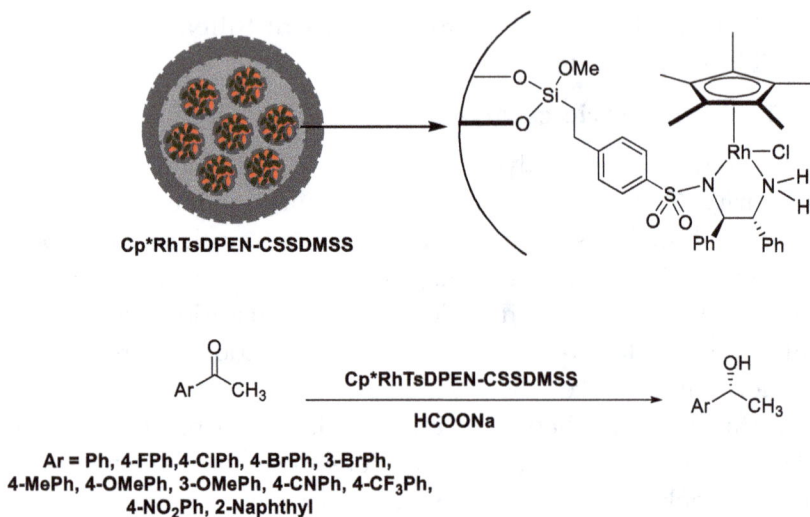

Scheme 7.1.   Asymmetric transfer hydrogenation of the aromatic ketone.

Cetyltrimethylammonium chloride (CTAC), the structure-directed template used in the synthesis, has the ability to function as a phase transfer catalyst. Hence, the incorporation of CTAC within CSSMSS presented a unique phase transfer catalyst in a biphasic catalysis system that greatly promoted the activity and enantioselectivity. Silica shell prohibited the leaching of chiral complex and thus was reused for 12 consecutive cycles without any appreciable decrease in enantioselectivity [12].

In another example, mesoporous silica nanospheres (MSNs) are used as supports for the immobilization of ruthenium catalyst, chiral $RuCl_2$-diphosphine-diamine complexes. Chiral $RuCl_2$-diphosphine-diamine complexes are important as they are air- and moisture-stable and can be readily purified by silica-gel chromatography [13]. The chiral ruthenium complex was immobilized on MSNs *via* a siloxy group installed in the diamine ligand (Scheme 7.2) [14]. Various chiral complexes (1–5) containing pendant siloxy were synthesized by treating $[\{RuCl_2(pcymene)\}_2]$ with chiral diphosphine in 1,2-dichloroethane/DMF at 80°C and subsequent treatment with siloxy-derivatized chiral 1,2-cyclohexanediamine. The catalytic potential of

(a)

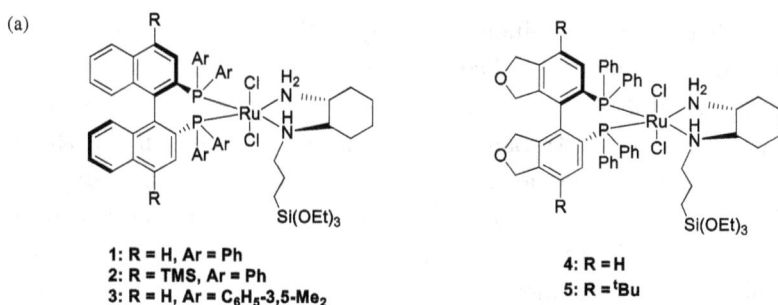

1: R = H, Ar = Ph
2: R = TMS, Ar = Ph
3: R = H, Ar = C$_6$H$_5$-3,5-Me$_2$

4: R = H
5: R = ^{t}Bu

(b)

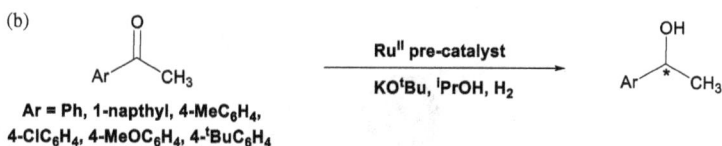

Ar = Ph, 1-napthyl, 4-MeC$_6$H$_4$,
4-ClC$_6$H$_4$, 4-MeOC$_6$H$_4$, 4-^{t}BuC$_6$H$_4$

(c)

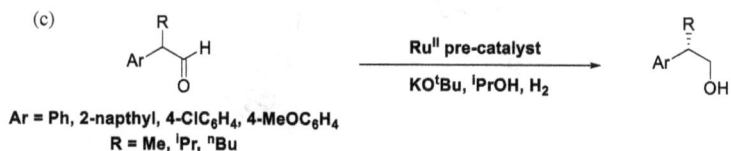

Ar = Ph, 2-napthyl, 4-ClC$_6$H$_4$, 4-MeOC$_6$H$_4$
R = Me, ^{i}Pr, ^{n}Bu

Scheme 7.2. (a) Structures of complexes 1–5, (b) asymmetric hydrogenation of aromatic ketones and (c) asymmetric hydrogenation of aryl aldehydes.

the heterogeneous ruthenium catalyst was explored in the asymmetric hydrogenation of aromatic ketones and racemic aryl aldehydes. The catalyst effectively converted aromatic ketones to the corresponding chiral secondary alcohols, while racemic aryl aldehydes afforded chiral primary alcohols with good enantioselectivities.

To further simplify the catalyst separation and effective recycling, silica-based magnetic nanocatalysts have been introduced, which combine the advantages of both mesoporous silica and magnetic nanoparticles (MNPs). Few examples have been reported in the literature for the immobilization of chiral catalysts over magnetic mesoporous silica. A report by Li and co-workers outlined a methodology for the immobilization of Ru-TsDPEN

(TsDPEN = N-(p-toluenesulfonyl)-1,2-diphenylethylenediamine) complex in a magnetic siliceous mesocellular foam (Scheme 7.3). The resulting catalyst demonstrated high activity and enantioselectivity in the asymmetric transfer hydrogenation of imine in the HCOOH-Et$_3$N system (Scheme 7.3(a)) and aromatic ketones in aqueous HCOONa (Scheme 7.3(b)). The fast and easy magnetic recovery of the catalyst allows convenient recycling of the expensive and air-sensitive metal complex [15].

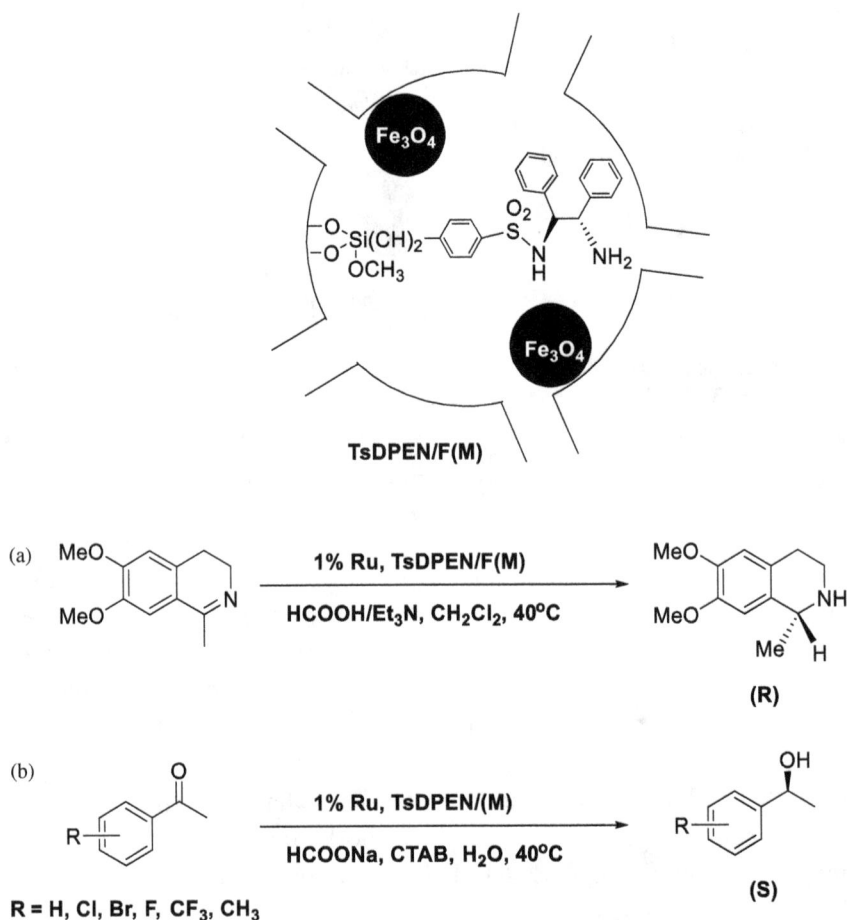

TsDPEN/F(M)

Scheme 7.3. Asymmetric transfer hydrogenation of (a) imines and (b) aromatic ketones.

Each enantiomer in racemic mixture has different pharmacological activity, so separation of enantiomers is of utmost importance in many industries and significant research has been done in this regard [16]. But, in spite of great accomplishments, enzymatic kinetic resolution of racemic mixtures still remains one of the biggest challenges since enzymatic activity gradually decreases. In this regard, immobilization of enzymes onto silica support could be the principal solution. Following this, *Candida rugosa* lipase immobilized on maghemite nanoparticles was synthesized by Gao *et al.* which showed good stereoselectivity in kinetic resolution of racemic carboxylates (Scheme 7.4). Enhanced stability over parent free enzyme, ease of recoverability and recyclability were the major outcomes of the protocol [17].

Chiral primary amine supported on MNPs was synthesized by Cheng *et al.* by treating silica-coated magnetic nanoparticles (SMNPs) with trimethoxysilane **A**, which was obtained using Pt-catalyzed hydrosilylation of the allylated precursors (Scheme 7.5) [18]. These were further used as an asymmetric bifunctional enamine catalyst in direct aldol reaction (Scheme 7.6). High activity and stereoselectivity, mild "on-water" conditions, easy

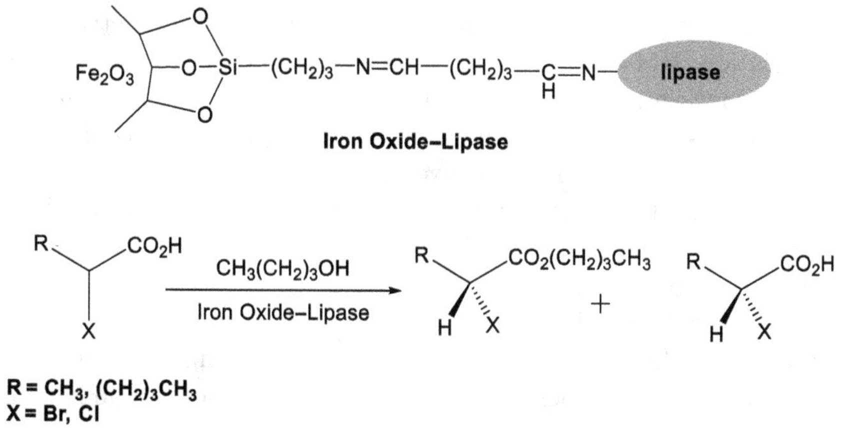

Iron Oxide–Lipase

R = CH₃, (CH₂)₃CH₃
X = Br, Cl

Scheme 7.4. Resolution of racemic carboxylates using *Candida rugosa* lipase immobilized on maghemite nanoparticles (iron oxide–lipase).

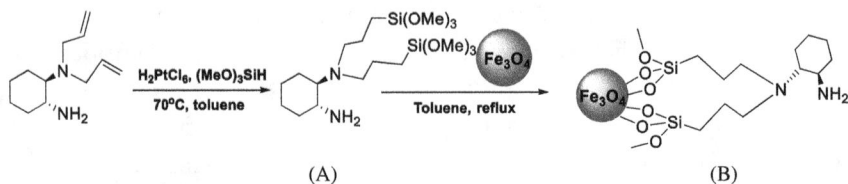

(A)   (B)

Scheme 7.5.   Synthesis of MNP-supported chiral primary amine catalyst.

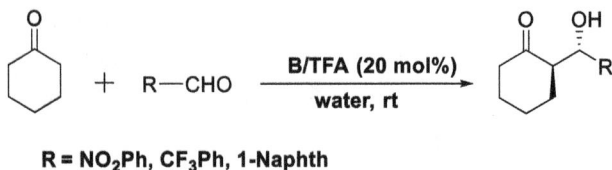

R = NO$_2$Ph, CF$_3$Ph, 1-Naphth

Scheme 7.6.   Asymmetric aldol reaction of cyclohexanone.

Note: TFA = Trifluoroacetic acid.

magnetic recovery and reusability up to 11 cycles were the green potentials of the protocol.

The Sharpless catalytic asymmetric dihydroxylation is of immense interest since it generates enantiomerically enriched vicinal diols [19]. Considering this, Kim and co-workers [20] fabricated magnetically separable hierarchically ordered mesocellular mesoporous silica support (HMMS) for highly effective catalytic asymmetric dihydroxylation. HMMS comprised mesocellular silica foam and ordered mesoporous walls. Moreover, MNPs were grafted on the pore surface of HMMS using thermal decomposition of iron propionate complex to make its recovery simple and convenient. The reported heterogeneous catalyst exhibited comparable rates and enantioselectivities as those observed with homogeneous ligand with the additional benefit of reusability up to eight cycles.

Another study by Reiser and co-workers reported two different azide-functionalized SMNPs for the immobilization of azabis(oxazoline) copper(II) complexes [21]. Azabisoxazoline ligands are potential candidates for the immobilization on heterogeneous support since they reduce the tendency of metal leaching from their complexes contrary to the corresponding metal-bisoxazoline

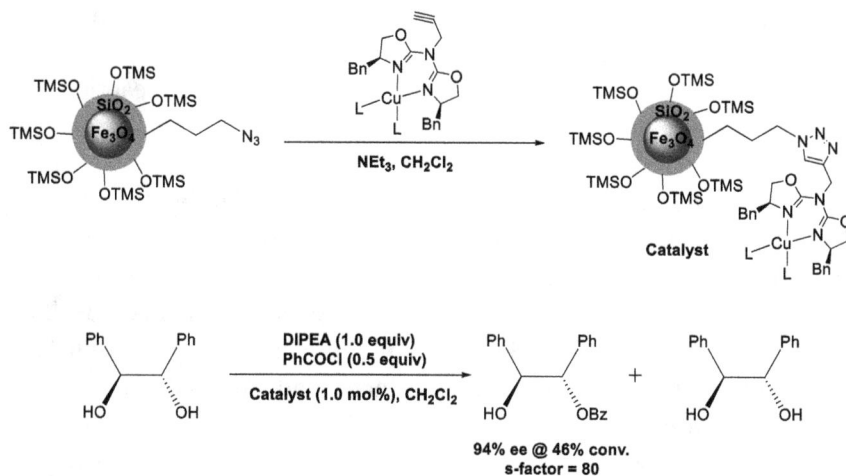

Scheme 7.7. Immobilization of azabis(oxazoline) copper(II) complexes over azide-functionalized SMNPs and its application in the asymmetric benzoylation of 1,2-diols.

complexes [22]. The catalyst was synthesized using azide linkage through click chemistry. Catalyst showed good activity and selectivity (up to 98% ee) for the asymmetric benzoylation of racemic 1,2-diols (Scheme 7.7). The selectivity was comparable to homogeneous catalyst (99% ee) and much more than common polymer resin supports, such as MeOPEG (up to 82% ee) or Merryfield resin (up to 67% ee). The catalyst was magnetically recovered and recycled for three times with no noticeable change in activity and selectivity [21].

7.2.2 One-pot cascade reactions

Catalyzed cascade reactions are one of the most promising approaches for the total synthesis of complex organic compounds [23]. They work on the blueprints of biosynthesis, avoid the use of expensive protecting group strategies and reduce the number of chemical steps, waste production and time, thereby increasing the synthetic efficiency. They can even be considered under the credentials of "Green Chemistry" [24]. Another distinguished advantage of such reactions is that a single catalyst is capable of promoting at least two reactions which are performed in a single operation and under the same reaction conditions.

Thus, considering the science and art of cascade reactions in total synthesis, Huang *et al.* fabricated two kinds of bi-functionalized mesoporous silica nanoparticles (MSNs) using both Brønsted acid (sulfonic acid) and Brønsted base (amine) [25]. The materials were synthesized by first co-condensing one of the two functional groups onto the internal channels. Subsequently, the second group was grafted post-synthetically on the external surface (Scheme 7.8). The so-formed MSNs — SAMSN-AP (sulfonic acid on internal surface and amine groups on external surface) and APMSN-SA (amine groups on internal surface and sulfonic acid on external surface) — were then utilized for the one-pot cascade reaction — sulfonic acid-catalyzed hydrolysis of 4-nitrobenzaldehyde dimethyl acetal and its further conversion using amine-catalyzed Henry reaction (Scheme 7.9).

Both SAMSN-AP and APMSN-AP led to 100% conversion of **A** and 97.7% and 98.1% yields of **C**, respectively. No conversion of

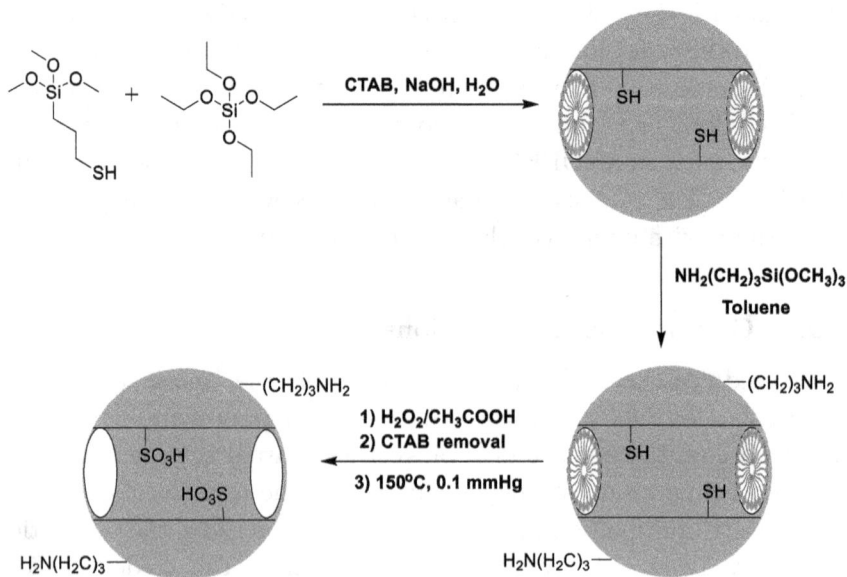

Scheme 7.8. Synthesis of bifunctional MSNs (SAMSN-AP).

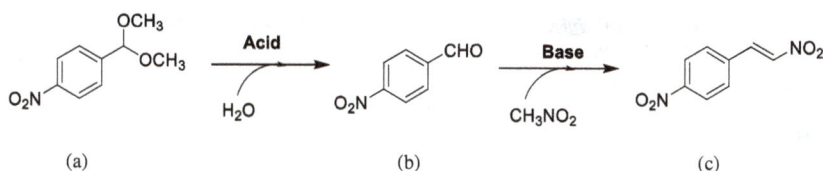

Scheme 7.9. One-pot cascade reaction of acid-catalyzed hydrolysis and base-catalyzed Henry reaction.

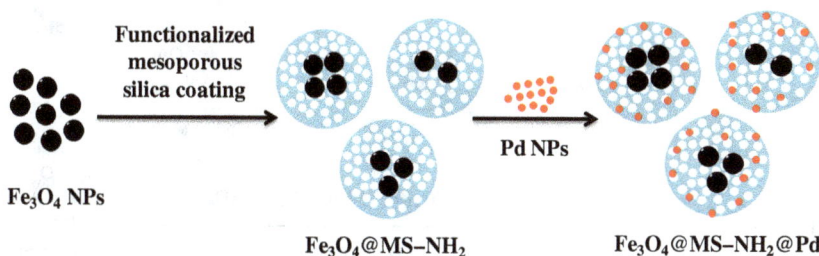

Scheme 7.10. Schematic illustration for the fabrication of core–shell satellite-structured Fe_3O_4@MS–NH$_2$@Pd

A into **C** was observed when either of their homogeneous analogs was used with functionalized MSN. Hence, the above example illustrated that immobilization of dual functionalities over SNPs, which are otherwise incompatible with each other in a homogeneous solution, facilitates their site separation and helps in controlling the cascade reactions independently.

Magnetic nanocomposites can also be combined with two different types of catalytically active species so as to utilize them in one-pot cascade reactions. In this regard, Li and co-workers fabricated core–shell satellite-structured Fe_3O_4@MS–NH$_2$@Pd in which Pd nanoparticles were dispersed on amine-functionalized MSNs (MS–NH$_2$) and Fe_3O_4 nanoparticles were scattered inside the silica (Scheme 7.10) [26].

MS–NH$_2$ served as a protecting shell for MNPs and provided coordination sites for the binding of Pd nanoparticles. Thus, the so-formed nanocomposites combined the catalytic properties of unbound amine groups for base-catalyzed reactions and noble metal Pd nanoparticles active sites. The integrated nanosystem was then

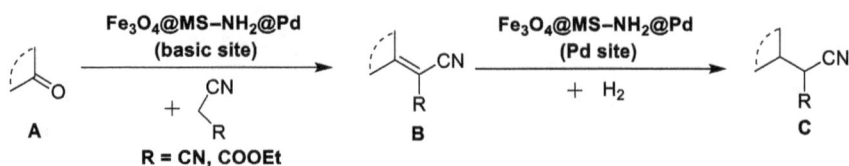

Scheme 7.11. One-pot Knoevenagel condensation–hydrogenation multistep cascade reaction using Fe_3O_4@MS–NH_2@Pd.

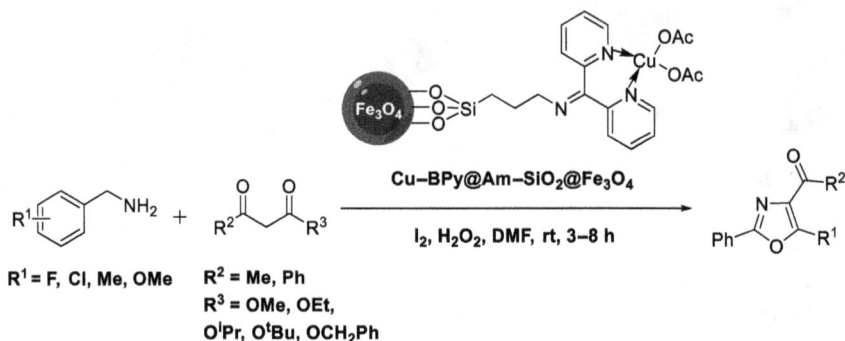

Scheme 7.12. Synthesis of polysubstituted oxazoles using Cu–BPy@Am–SiO_2@ Fe_3O_4.

employed for the one-pot Knoevenagel condensation–hydrogenation multistep cascade reaction (Scheme 7.11).

The desired products, α-alkylated nitriles, were obtained under mild reaction conditions and in a single vessel. Besides, the MNPs provided easy recoverability and recyclability up to four runs.

Similarly, Sharma and co-workers have designed a silica-coated magnetic copper nanocatalyst (Cu–BPy@Am–SiO_2@Fe_3O_4) for the tandem oxidative cyclization reaction to obtain biologically active polysubstituted oxazoles (Scheme 7.12) [27]. Advantages such as green reaction conditions, high atom economy, excellent turnover numbers, magnetic recoverability, recyclability up to eight consecutive runs and applicability to gram-scale synthesis present a sustainable approach for oxazole synthesis.

7.2.3 Photocatalytic degradation of dyes

Synthetic dyes are important constituents of plentiful industries, such as textile, leather tanning, food technology, paint and varnishes, hair colorings and many more [28]. It is estimated that more than 10,000 dyes are commercially available and approximately 7 lakh tons of dyes are produced annually in the world [29]. Discharge of industrial effluents containing even very low concentration of dyes in water bodies is highly visible, undesirable and causes considerable environmental pollution. Besides, dyes are found to possess severe detrimental health hazards due to their toxicity. Several physical, chemical and biological methods have been reported for their removal. Among all, photocatalytic degradation has gained enormous interest in the treatment of wastewater containing synthetic dyes [30]. Such processes involve the use of a semiconducting material which upon absorption of an activated photon, having energy equal to or greater than the band gap of the semiconductor, creates electron–hole pairs leading to light-induced redox processes. They provide several key advantages, such as low cost, mild degradation conditions, no waste disposal problem and complete mineralization.

Various semiconductors are known for photocatalytic processes, such as TiO_2, ZnO, Fe_2O_3, CdS and ZnS. Among all, TiO_2 is one of the favored materials due to its chemical and photocatalytic stability, biocompatibility and superior optical and electrical properties [31]. It has four mineral forms, namely, titanium dioxide, anatase, rutile and brookite. Among all, anatase is mainly used as a photocatalyst under UV irradiation. Detailed mechanism for the photocatalytic dye degradation is shown in Figure 7.1 [32]. Much research has been done on the use of aqueous TiO_2 suspensions for photocatalytic degradation; nevertheless, issues such as separation and recyclability have been scarcely addressed. Thus, immobilization of TiO_2 provides an appropriate solution toward this problem. Support plays a key role in designing a photocatalyst as immobilization significantly reduces the active surface area of photocatalysts and hence decreases the photocatalytic activity. Therefore, it is necessary to develop a

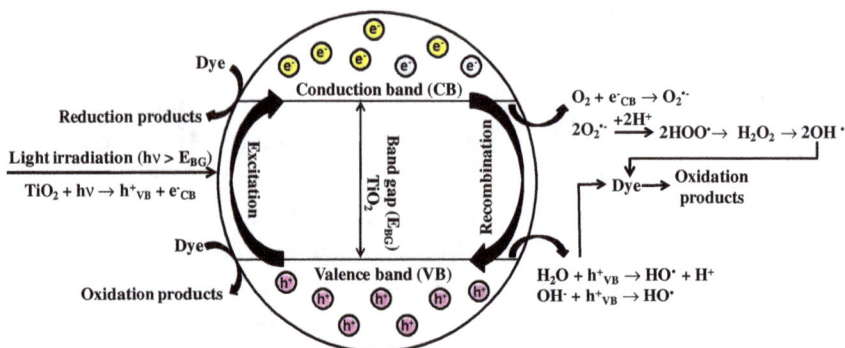

Figure 7.1. Mechanism for photocatalytic dye degradation.

multifunctional photocatalytic system which provides high substrate adsorption capacity, good light harvest properties, accessible active sites and recyclability.

Recently, Singh *et al.* designed a photocatalyst by coating TiO_2 on fibrous nanosilica (KCC-1) using atomic layer deposition technique followed by hydrolysis and condensation to develop TiO_2 coating on KCC-1 fibers [33]. The so-formed KCC-1/TiO_2 nanocatalyst was then utilized for the photocatalytic degradation of Rhodamine-B, methylene blue and phenol (Scheme 7.13). It is reported that the fibers of nanosilica adsorb the dye molecules, which allows easy interaction of dyes with TiO_2 active sites. This reduces the distance for the migration of photogenerated oxidizing species (excitons) and the dye molecules, which in turn is responsible for the enhanced photocatalytic activity of KCC-1/TiO_2 than the well-known MCM-41- and SBA-15-supported TiO_2. High TiO_2 loading, better accessibility of active sites, high surface area and enhanced light-harvesting properties are some of the other advantages provided by KCC-1/TiO_2 which accelerated the rate of degradation.

In addition to nanosilica, MNPs can also serve as a useful support material enabling efficient immobilization and facile magnetic recovery of the photocatalysts. However, Li *et al.* recently reported that direct deposition of TiO_2 on magnetic iron oxides results in photodissolution of Fe_3O_4 and electron–hole recombination [34]. Thus, in order to overcome these problems, researchers introduced an

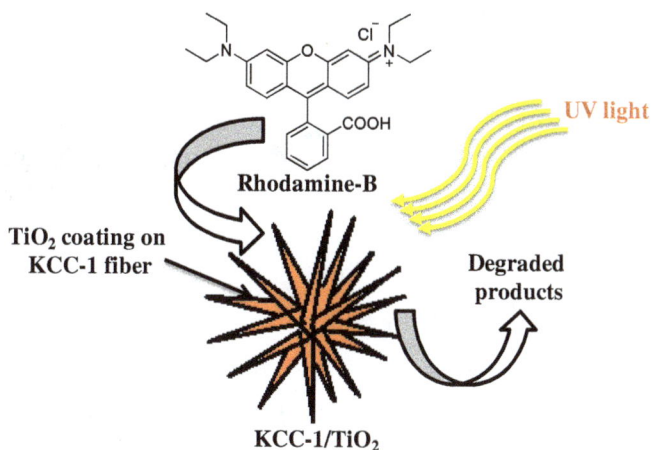

Scheme 7.13. Rhodamine-B degradation using KCC-1/TiO$_2$.

intermediate silica layer between the iron oxide support and TiO$_2$. Besides screening the charge exchange between TiO$_2$ and the iron support, the porous silica coating provides large surface area-to-volume ratio for enhanced catalytic activity. In this regard, Greene *et al.* synthesized CoFe$_2$O$_4$@SiO$_2$@TiO$_2$ magnetic nanocatalyst for the photooxidation of methylene blue using UV light [35].

In order to strongly immobilize titania crystals on magnetic support, Cheng *et al.* modified the surface of magnetic/silica using polyacrylic acid (PAA) [36]. PAA formed strong covalent bonds between TiO$_2$ nanocrystals and SiO$_2$, and thereby provided strong interface binding of TiO$_2$. Further, calcination was done at 500°C which converted amorphous TiO$_2$ into crystalline anatase TiO$_2$. Scheme 7.14 shows the preparation process for magnetic/SiO$_2$/TiO$_2$.

Further, the nanocomposite was utilized in the degradation of Rhodamine B under UV irradiation. Intensity of the peak at $\lambda_{\text{max}} = 553$ nm decreased after 105 min which indicated the photodegradation of the dye by the catalyst.

Recently, noble metal nanoparticles (especially Au and Ag) have been recognized as a new class of materials for harvesting solar energy for chemical transformations [37]. Intensive studies have been carried out to increase the optical absorption of photocatalysts through

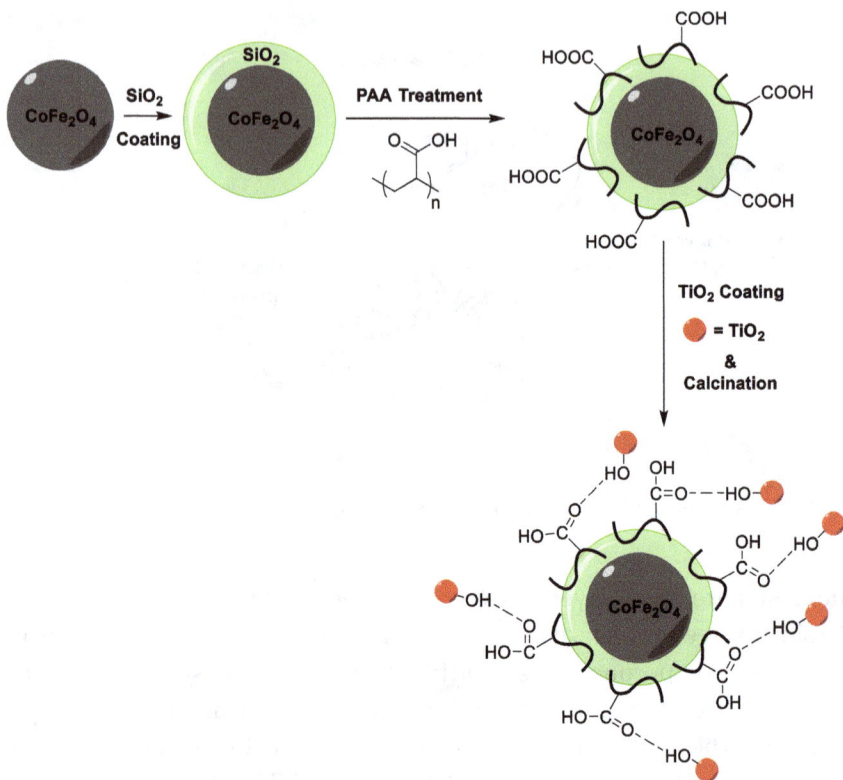

Scheme 7.14.   Schematic illustration for the preparation of magnetic/SiO_2/TiO_2.

either narrowing the band gaps of semiconductors or inserting light-absorbing components into semiconductors with a wide energy gap. On this note, partial reduction of silver ions in silver halides has been performed to fabricate metallic Ag-doped silver halide [AgX(Ag)] [38]. The presence of Ag^0 in AgX significantly improves the photocatalytic activity by adding extra energy states between the conduction and valence band of AgX, thus enabling strong absorption of visible light. Surface plasmon resonance of Ag^0 nanodomains also enhances the optical absorption. Besides, the high electron density in Ag^0 increases the chemical stability of AgX by diminishing the photoreduction of AgX during photocatalysis. But, aggregation of AgX(Ag) nanoparticles often hampers their practical utilization.

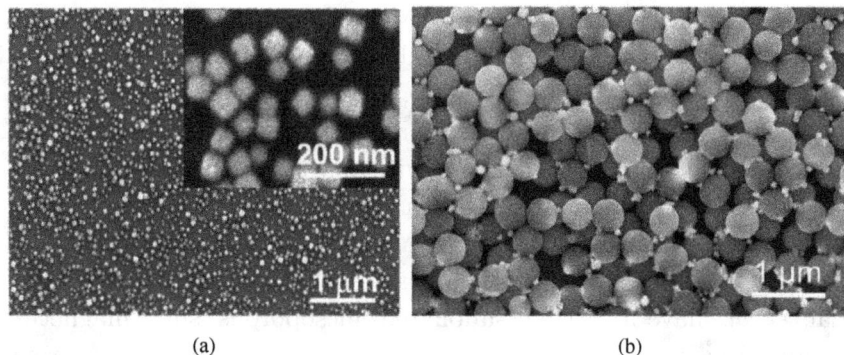

Figure 7.2. SEM images of (a) AgCl(Ag) nanocubes, (b) SiO$_x$/AgCl(Ag) hybrid nano structure (adapted from Ref. [39]).

In this regard, recently, Rasamani and co-workers immobilized AgCl(Ag) nanocubes ($\sim$60 nm) onto silica nanospheres (SNSs) with hundreds of nanometer in size [39]. For this, first, the SNSs were synthesized using Stöber method and modified using APTES for surface functionalization. Further, AgCl nanocubes were synthesized and illuminated with a Xenon lamp to partially convert AgCl to metallic Ag. The resulting AgCl(Ag) nanocubes (Figure 7.2(a)) were mixed with APTES-modified SNSs to finally obtain SiO$_x$/AgCl(Ag) (Figure 7.2(b)). The so-formed nanocomposites were then utilized for the photocatalytic degradation of methylene blue dye.

7.2.4 Biomass conversion

Production of fuels and chemicals from renewable resources is gaining tremendous attention due to the continuously dwindling fossil fuel reserves and increase in global warming. Renewable biomass, the only carbon-containing resource, are the most promising feedstocks for the development of alternative sources of energy and chemical intermediates. Biomass comprises four major components, cellulose, hemicellulose, starch and lignin, which on hydrolysis gives the raw materials, i.e., sugars [40].

For this, the enzymatic degradation of cellulose is considered as one of the most effective ways to obtain sugars, as high selectivity is achieved under moderate hydrolysis conditions. However, owing to the complex structure of lignocellulosic materials, it is difficult for enzymes to access cellulose, hence, pretreatment of lignocellulosic biomass is required [41]. Acid hydrolysis is another process which is economical but is associated with several drawbacks, such as environmental unfriendliness, harsh reaction conditions and tedious separation of products. To overcome these issues, immobilization of acid on silica support could be an ideal solution [42]. For instance, Lai *et al.* have reported sulfonated mesoporous silica-magnetite nanocomposites [41]. The material comprised sulfonated mesoporous silica as active hydrolysis catalyst and magnetic iron oxide particles that facilitate catalyst separation from reaction media. The catalyst effectively hydrolyzed β-1,4-glucan, to produce glucose in 96% and 50% yield from cellobiose and amorphous cellulose, respectively. Besides, it also displayed good hydrothermal stability and recyclability and is environmentally benign in contrast to the conventional solid acids.

Another group also used a similar type of catalyst for hydrolyzing cellulose into reducing sugar in ionic liquid (IL) (Scheme 7.15) [43].

These sugars can be further converted into useful chemicals including sugar alcohols, γ-valerolactone, and 5-hydroxymethylfurfural (HMF) for biofuel production.

5-Hydroxymethylfurfural, a key platform molecule, is generated by the dehydration of C6-based carbohydrates. It is a versatile precursor for the formation of high-energy content fuels, chemicals and polymers. HMF upon oxidation generates important chemicals such as 2,5-furandicarboxylic acid (FDCA) and 2,5-diformylfuran (DFF) [44]. DFF is an important potential chemical intermediate as it can be employed in the synthesis of furanic polymers, pharmaceuticals, antifungal agents and renewable furan-urea resin [45].

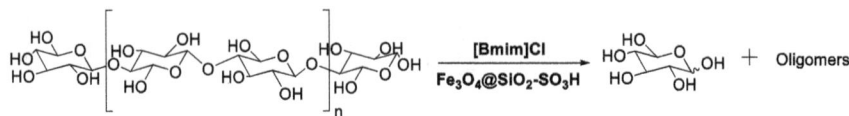

Scheme 7.15. Hydrolysis of cellulose to reducing sugar in IL.

Following this, Zhang and co-workers have fabricated magnetic-supported Ru(III) catalyst for the aerobic oxidation of HMF to DFF. Amino-modified SMNPs were formed and further treated with an aqueous solution of $RuCl_3$ to obtain $Fe_3O_4@SiO_2-NH_2-$Ru(III) catalyst [46]. The loading of ruthenium in the catalyst was determined by inductively coupled plasma atomic emission spectroscopy to be 1.4 wt.%. The catalyst was further investigated in aerobic oxidation of HMF to DFF (Scheme 7.16). After 4 h, a high

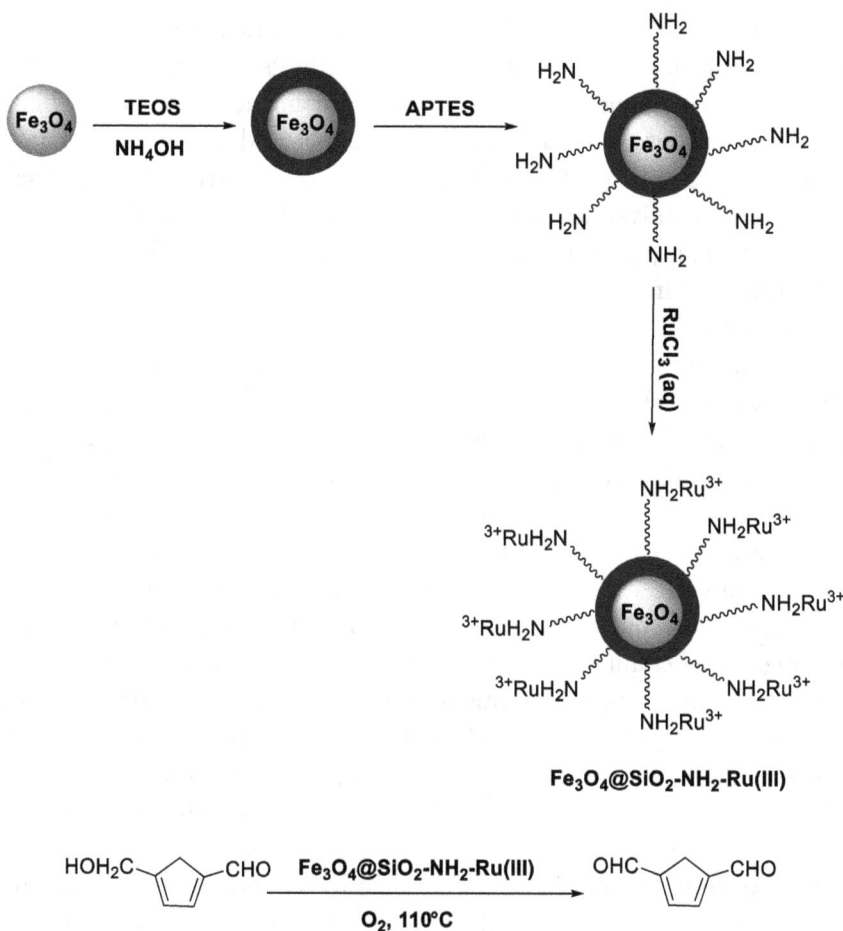

Scheme 7.16. Synthesis of $Fe_3O_4@SiO_2-NH_2-Ru(III)$ catalyst and its application in aerobic oxidation of HMF.

DFF yield of 86.4% and HMF conversion of 99.3% was obtained at 120°C, while excellent HMF conversion of 99.7% and DFF yield of 86.8% was achieved after 16 h in air. High catalytic performance in air, facile recovery and recycling make the reported protocol economical and convenient.

Another group utilized ruthenium(III)-functionalized SMNPs as catalyst for the synthesis of succinic acid from levulinic acid [47]. Succinic acid and its salts are the key building blocks for a variety of products and chemical intermediates in food, fine chemicals and pharmaceutical industries [48]. Therefore, various thermochemical pathways have been developed for the synthesis of succinic acid [49], but most of them suffer from harsh conditions, such as high temperature, use of harsh and toxic acids and oxidants and low yields. In this respect, the exploitation of Ru(III)-functionalized SMNPs as catalyst for levulinic acid oxidation has established an excellent example of green aqueous-phase catalytic oxidation [47]. The protocol involved the use of molecular oxygen as oxidant and stable magnetically recoverable catalyst. Moreover, it circumvents the need for any base and organic solvent. The only associated drawback was the low solubility of molecular oxygen in water, thus requiring high pressure (10 bar).

Recently ILs are gaining much attention as both a novel reaction medium and as a catalyst for the various organic transformations. However, the use of large amount of ILs as homogeneous catalyst is associated with serious environmental and economic concerns. Hence, heterogeneous solid-supported ILs seem more promising than their homogeneous counterpart. In this context, Coutinho and co-workers have developed supported ionic liquid NPs (SILnPs) by immobilizing IL, 1-(tri-ethoxy silyl-propyl)-3-methyl-imidazolium hydrogen sulfate (IL-HSO$_4$), on the surface of SNPs through covalent bonding between IL cation and silica surface (Scheme 7.17) [50]. The potential of SILnPs was tested for the catalytic dehydration of fructose to the HMF. Within 30 min, SILnPs were able to convert 99.9% fructose to the desired HMF (63% yield) at 130°C. Moreover, the catalyst was effortlessly recycled seven times without noticeable loss in its activity and selectivity.

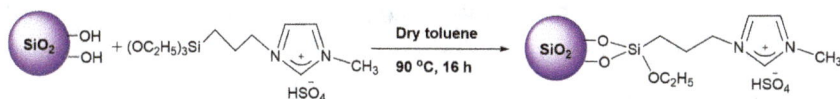

Scheme 7.17. Immobilization of IL, 1-(tri-ethoxy silyl-propyl)-3-methyl-imidazolium hydrogen sulfate, on SNPs.

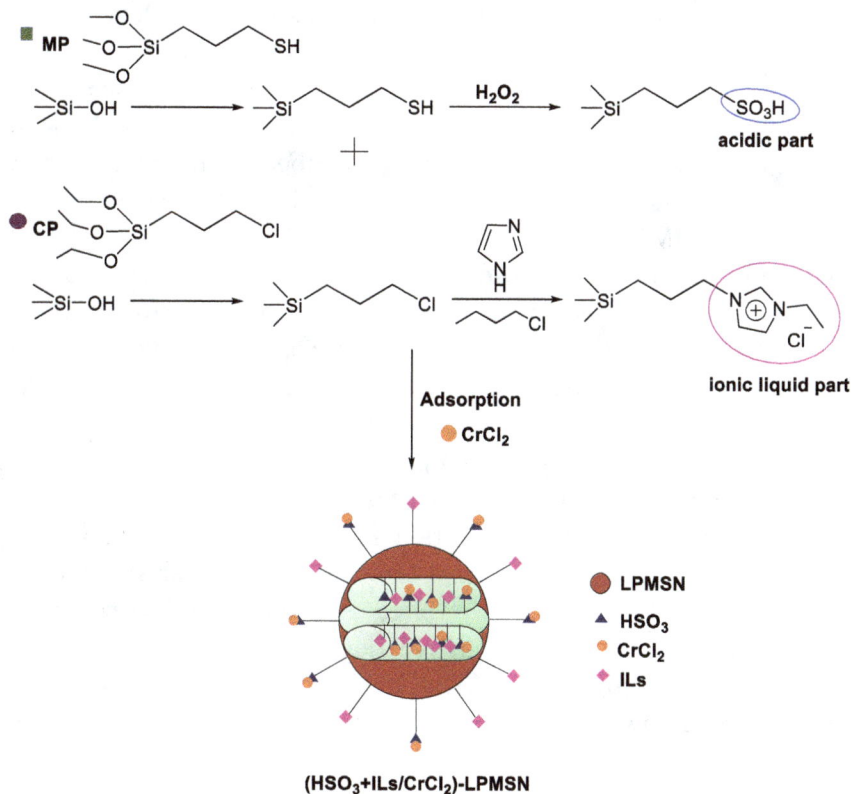

Scheme 7.18. Preparation of the bi-functionalized MSN.

Another group reported the conversion of fructose to HMF using sulfonic acid and IL bi-functionalized mesoporous silica NPs (MSNs). Synthetic procedure for catalyst preparation has been depicted in Scheme 7.18. The fructose conversion was achieved under milder conditions (90°C and 3 h) and HMF was produced with the highest yield of 72.5%. The high catalytic activity of the bi-functionalized MSN catalyst is attributed to the co-existence of functional groups of HSO_3 acid and [EMIM]Cl/$CrCl_2$ IL. It was reported that apart from sulfonic acid groups, chloride ions also contribute to the catalytic activity owing to their nucleophilicity, and the acidic C-2 proton of the imidazolium ring of IL also promotes the fructose dehydration. The authors also conducted kinetic studies which suggested that bi-functionalized MSN reduces the activation energy and thus accelerates the dehydration of fructose. Overall, the catalytic system is economical, selective and can be easily operated at large scales [51].

In another report, Peng *et al.* [52] described a strategy for direct one-pot conversion of cellulose to HMF. Generally, conversion of cellulose to HMF involves three main reactions: (i) acid-mediated depolymerization of cellobiose to glucose, (ii) base-catalyzed isomerization of glucose to fructose and (iii) dehydration of fructose to HMF in the presence of a catalyst (Scheme 7.19). So, the authors fabricated acid–base bi-functionalized, large-pored mesoporous SNPs (LPMSN) for cooperative one-pot conversion of cellulose. The LPMSN was synthesized using Brij-97 as the template and dimethyl-*o*-phthalate as the swelling agent which was further functionalized with acid (SO_3H) and/or base (NH_2) group to obtain LPMSN–SO_3H, LPMSN–NH_2 and LPMSN-both. To appreciate the role of the catalyst, glucose, fructose and cellobiose were also tested as reactants besides cellulose. The authors demonstrated that reactions that need acid and base are well-catalyzed by LPMSN–SO_3H and LPMSN–NH_2, respectively, while LPMSN-both is beneficial for direct one-pot conversion of cellulose to HMF.

7.2.5 Biocatalysis

Biocatalysis can be defined as the utilization of natural substances such as enzymes or cells for speeding up/catalyzing chemical reactions especially those having high industrial significance. In fact, the term "biocatalysis" is not new and has been around for quite a long time. According to the literature reports, biocatalysis underpins some of the oldest chemical transformations which include production of

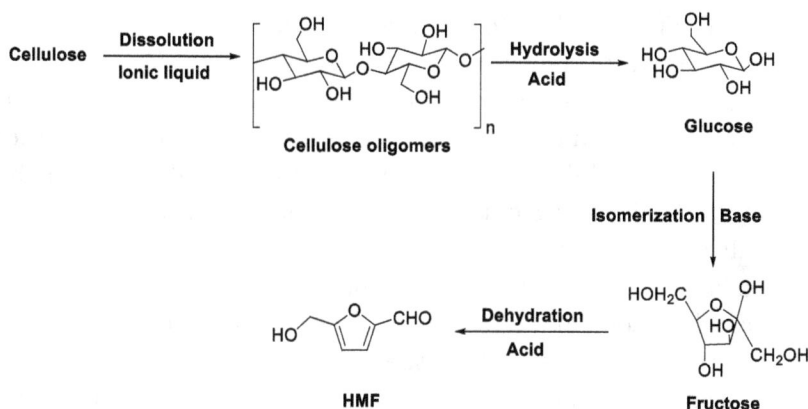

Scheme 7.19. General schematic illustration for the formation of HMF converted from cellulose through a series of reactions.

alcohol *via* fermentation and cheese *via* enzymatic breakdown of milk proteins [53]. Here, it becomes imperative to highlight that although the fermentation technology had been frequently incorporated for the development of blockbuster drugs, e.g. penicillin, industries kept favoring transition metal catalysts instead of the biocatalysts for quite some time until the late 20th century or the beginning of 21st century [54]. This was because of the fact that under the industrial synthetic conditions, natural enzymes invariably tended to be unstable and also gave very poor yields. But thankfully, technology caught up soon and it was the incredible advancements in gene engineering and enzymatic evolution that biocatalysis was welcomed with open arms.

Today, biocatalysts are being employed extensively by big pharmaceutical companies which have integrated this green technology with traditional medicinal chemistry for synthesizing small-molecule drugs [55]. The expansion in this use is quite obvious considering the enormous benefits they offer which include being environmentally friendly and possesing excellent enantio-, regio- and chemo-selectivities. Undeniably, it is the advancement in basic understanding of the protein structure–function relationship that has enabled scientists to carry out the rational design of different biocatalytic systems whose properties can be tailored/routinely engineered in the laboratory. Many of the designed biocatalysts are being used in redox reactions, transesterification processes, enantioselective synthesis, etc., but factors such as exorbitant cost, poor availability and difficulties in recovery and recycling limit their large-scale applicability. Thus, in this direction, efforts are being directed toward finding out alternatives for free/crude enzymes in biocatalyzed reactions.

Among different methodologies reported that include entrapment of biocatalysts on porous materials (such as ion-exchange resins), the adsorption of enzymes on the surface of a material and the immobilization of enzymes on different supports *via* functionalization and/or covalent binding [56] (Figure 7.3), the covalent immobilization method using suitable support materials to enhance the reusability properties of enzymatic systems has gained prominence. SiO_2 and $SiO_2@Fe_3O_4$ NPs have emerged as very good support matrices for the immobilization of enzymes [57].

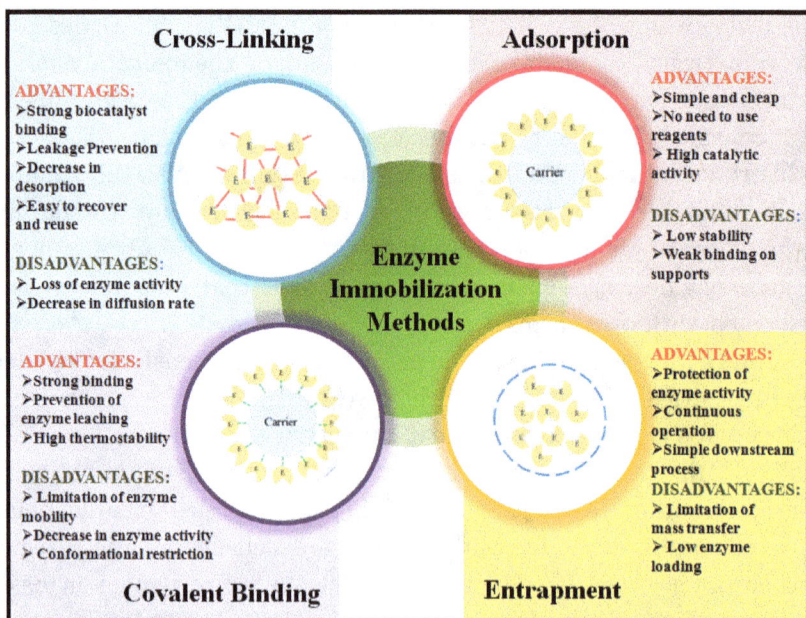

Figure 7.3. Different enzyme immobilization techniques: their advantages and disadvantages.

The magnetically separable NPs have the additional advantage of stability and easy separation and hence the encapsulation of enzymes into magnetic materials has particularly emerged as the approach of choice. It was Reetz *et al.* who reported for the first time the entrapment of enzymes in magnetite-containing sol–gel materials for heterogeneous biocatalysis in the year 1998 [58]. A previously reported strategy was utilized for simultaneous entrapment of lipases like *Pseudomonas cepacia* and *Candida antarctica* and the nanostructured superparamagnetic magnetite into hydrophobic sol–gel materials derived from methyl/ethyl/propyl/butyltrimethoxyorthosilane (R-TMOS). Subsequently, the activity of the supported biocatalyst was investigated in the esterification reaction (esterification of lauric acid with *n*-octanol). The results revealed that lipase-/magnetite-containing gels exhibited three times enhancement in its activity as compared to the non-entrapped free enzymes. This protocol displayed an additional advantage of facile separation *via* the use of an external magnet.

Lee and co-workers fabricated discrete, uniformly sized SMNPs (magnetite@silica) *via* a simple one-pot process using reverse micelles as nanoreactors [59]. Furthermore, they carried out the functionalization of the synthesized magnetite@ SNPs by mixing TEOS with silane agents such as 3-aminopropyltriethoxysilane. Thereafter, for testing the biological applicability of the developed nanomaterial, enzyme molecules (lipase and α-chymotrypsin) were cross-linked to form clusters on the surface of these functionalized NPs. From the results of the various catalytic activity tests, it was found that the resulting hybrid cross-linked enzyme cluster (CEC) magnetite composites were magnetically separable, were highly active and stable under harsh conditions and could be recycled for multiple runs.

The simultaneous entrapment and/or immobilization of biological macromolecules like enzymes and/or magnetite/magnetic carbons in sol–gel materials has emerged as an attractive strategy to produce bioactive, mechanically stable and magnetically separable materials. Using a similar kind of approach, Yang *et al.* prepared biocatalysts by entrapment of horseradish peroxidase (HRP) onto magnetite-containing spherical SNPs [60]. A reverse-micelle technique was

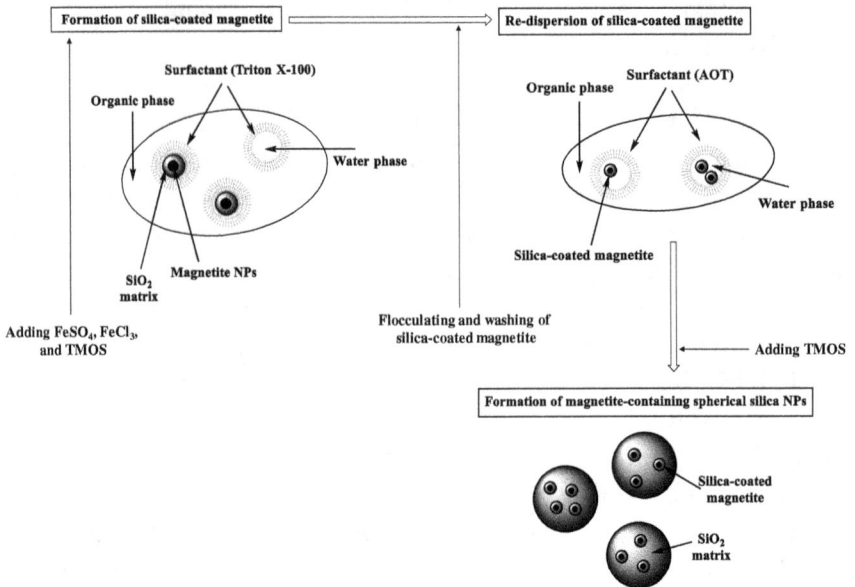

Scheme 7.20. Preparation of enzyme entrapped on magnetite-containing spherical SNPs.

utilized for synthesizing this hybrid material (Scheme 7.20). After the prehydrolysis of the silica precursor, the enzyme HRP was simply encapsulated in the magnetite-containing silica material by mixing and stirring a diluted solution of the protein in a phosphate buffer. The results of the experiments revealed that the maximum entrapment efficiency was 85–90%, and HRP did not leach from the surface of the support even after 60 days which indicated the excellent stability of the biocatalyst. Additionally, this material also showed excellent prospects of being employed in a novel magnetic separation immunoassay for the quantitative determination of gentamicin which is a well-known antibiotic used to treat several types of bacterial infections. The detection limit for gentamicin was found to be 160 ng mL^{-1}.

The entrapment approach has also been successfully exploited by Kuwahara *et al.* for synthesizing an efficient new type of heterogeneous biocatalyst [61]. As depicted in Figure 7.4, in this biocatalytic system, enzymes (*Candida antarctica Lipase A* (CALA)) are directly

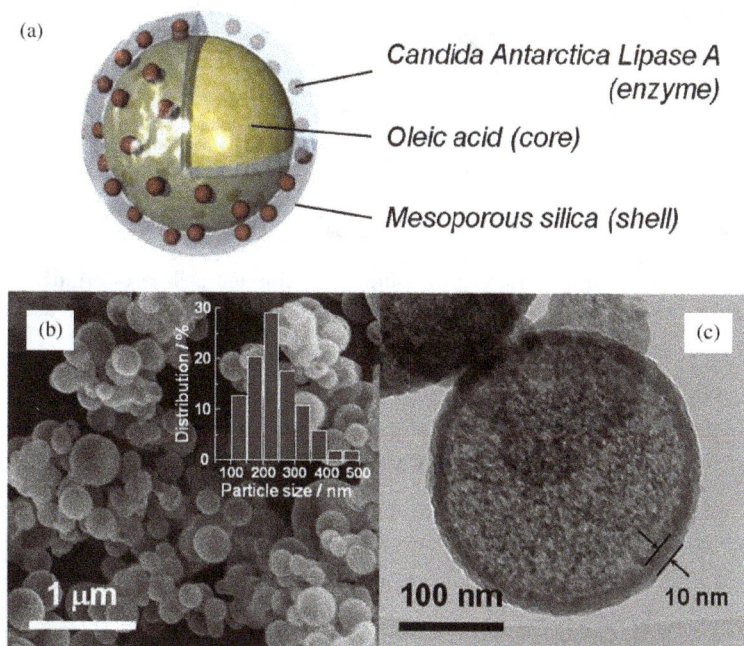

Figure 7.4. (a) Illustration of structure, (b) SEM micrograph (inset shows a particle size distribution diagram) and (c) TEM micrograph of CALA@OSN (adapted from Ref. [61]).

entrapped within spherical SNPs that comprise an oil-filled core while the shell material is made up of oil-induced mesoporous silica shell. An oil-in-water emulsion templating method was employed for synthesizing the CALA-embedded oil-filled SNPs (CALA@OSN). The enzymes entrapped within such a kind of architecture not only exhibited exceptional catalytic performance in both water as well as organic media but also showed good stability and recyclability [61].

Recently, enzyme-assisted cellulosic conversion has drawn the research interest of the scientific fraternity as an alternative green approach to produce useful chemicals/biofuels that reduces experimental costs, inhibits the generation of unwanted by-products and elevates reaction efficiency and specificity. In this context, Lee *et al.* reported the fabrication and application of different

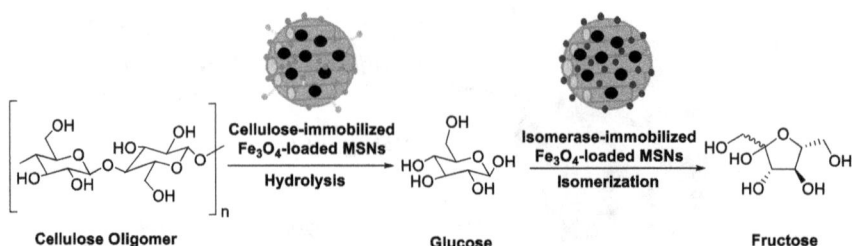

Scheme 7.21. An illustration expressing a continuous cellulose-to-glucose and glucose-to-fructose conversion sequence by using cellulase and isomerase separately immobilized onto Fe_3O_4-loaded MSN catalysts.

enzyme-immobilized MSNs (cellulase-immobilized and isomerase-immobilized Fe_3O_4-loaded MSNs) for the multistep cellulose-to-fructose conversion in an aqueous solution (Scheme 7.21) [62]. Using this catalytic route, fructose was obtained in approximately 51% yield which is by far the maximum yield obtained in industry for the production of fructose from glucose.

For carrying out the hydrolysis of cellobiose, Califano *et al.* developed a biocatalyst (BG/WSNs) by immobilizing β-glucosidase enzyme on wrinkled silica nanoparticles (WSNs) *via* physical adsorption (Scheme 7.22) [63]. The reason for choosing WSNs as the support matrix has been illustrated in Figure 7.5.

The synthesized matrix exhibited both a central–radial pore structure as well as a hierarchical trimodal micro/mesoporous pore size distribution. From the TEM image of WSNs (Figure 7.6), it could be seen that these are colloidal particles whose internal morphology primarily comprises silica fibers or wrinkles which spread homogeneously in all directions, forming central–radial pores extending radially outward. After BG adsorption, there is a marked change in morphology, although the diameters remain similar. The TEM image of BG/WSNs show that the dark region is much more extended in this case that indicates a high density of wrinkle in this region which is due to the presence of this immobilized protein. Results of the kinetic studies revealed that the activity of the immobilized enzyme was almost similar to that of the free enzyme. It was inferred that the immobilization procedure employed was efficient enough to preserve

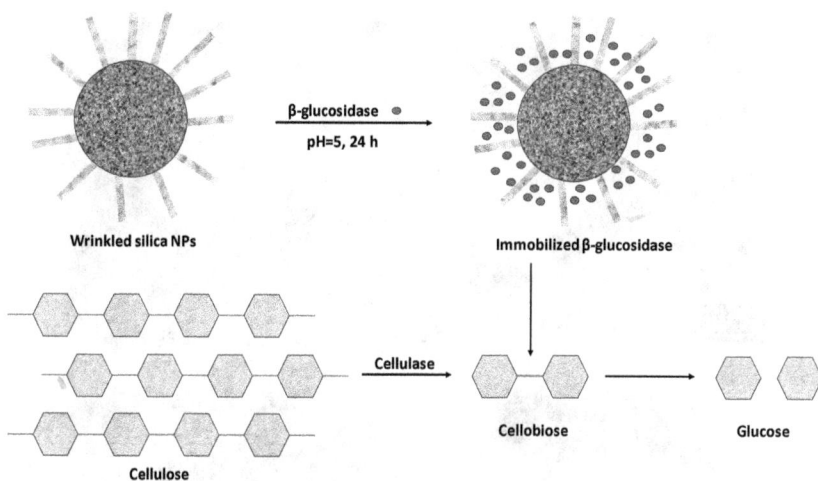

Scheme 7.22. Fabrication of biocatalyst.

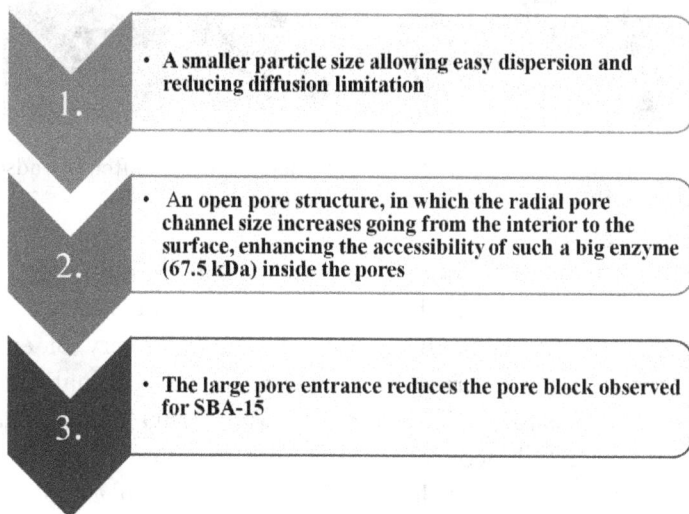

1.
- A smaller particle size allowing easy dispersion and reducing diffusion limitation

2.
- An open pore structure, in which the radial pore channel size increases going from the interior to the surface, enhancing the accessibility of such a big enzyme (67.5 kDa) inside the pores

3.
- The large pore entrance reduces the pore block observed for SBA-15

Figure 7.5. Advantages of MSNs as the support matrix.

most of the secondary structure of the enzyme and thus its catalytic activity.

From the biotechnological perspective, extremophilic enzymes show great potential to be used in various substrate transformations

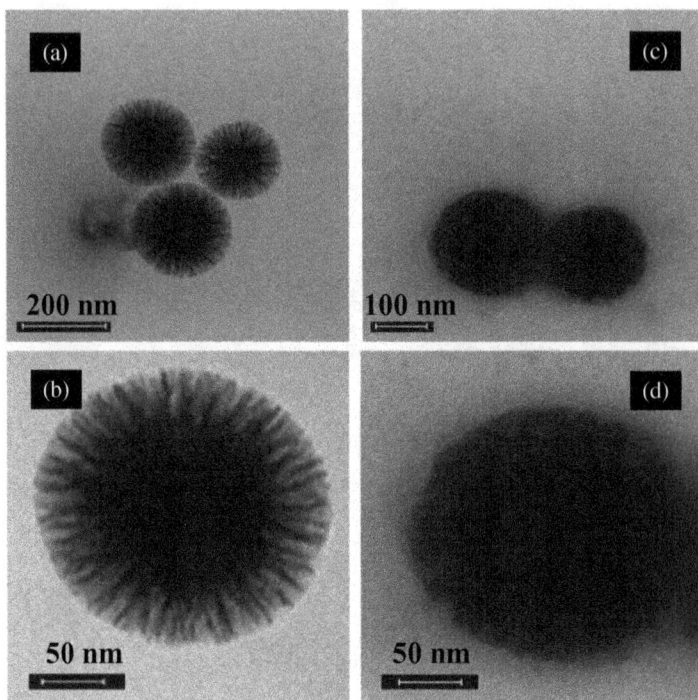

Figure 7.6. TEM images of WSNs: (a), (b) before, (c), (d) after BG adsorption (adapted from Ref. [63]).

under non-conventional harsh conditions, not achievable with normal enzymes; MNP-based immobilization of enzymes offers great advantages including excellent dispersibility, facile recovery of enzymes from the reaction mixture *via* application of an external magnetic field and good reusability. This impelled Sommaruga *et al.* to examine the effect of MNP bioconjugation on the catalytic properties of a thermostable carboxypeptidase derived from *Sulfolobus solfataricus* (CPSso) [64]. For this, first, they prepared high-quality water-soluble iron oxide nanoparticles in organic solvents which were coated with a thick silica shell *via* the reaction with tetraethoxysilane (TEOS) in NH_4OH, and thereafter, these silica-encapsulated iron oxide nanoparticles were reacted with the chelating agents ICPTES-NTA and $NiCl_2$. This led to the production of an Ni-functionalized nanoparticle (NiNTASiMNP), as described in Scheme 7.23. Finally,

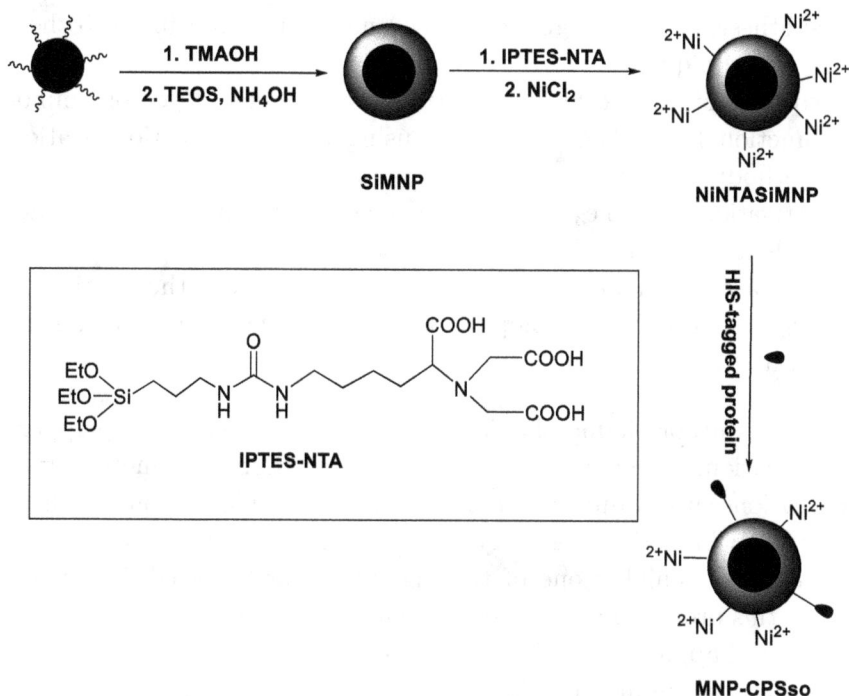

Scheme 7.23. Procedure of CPSso immobilization on NiNTASiMNP.

the CPSso was immobilized onto the surface of the silica-coated iron oxide nanoparticles through NiNTA-His tag site-directed conjugation. These enzyme-loaded MNPs could be employed as an efficient biocatalyst for the synthesis of N-blocked amino acids.

Working on similar lines, Cruz *et al.* carried out the immobilization of enzymes *S. Carlsberg* and *Candida antarctica lipase B* (CALB) on fumed SNPs for application in non-aqueous media and found that the catalytic activities of these immobilized biocatalytic systems were remarkably high [65].

Also, cyclodextrin glucanotransferase enzyme was immobilized on aminopropyl-functionalized silica-coated superparamagnetic nanoparticles and thereafter the properties of the immobilized enzyme were scrutinized. The synthesis of the enzyme-immobilized NPs was accomplished in a stepwise manner, as described in the following:

- synthesis of core magnetic (Fe_3O_4) nanoparticles using solvothermal technique;
- coating of the Fe_3O_4 nanoparticles by a dense layer of amino-functionalized silica (NH_2–SiO_2) using an *in situ* functionalization method;
- activation of the $Fe_3O_4@NH_2$–SiO_2 nanoparticles using gluteraldehyde as the bifunctional cross-linker;
- CGTase immobilization on the surface of the activated $Fe_3O_4@NH_2$–SiO_2 nanoparticles *via* a covalent attachment strategy.

The authors found that the approach employed for (i) support preparation, (ii) activation and (iii) optimization of immobilization conditions successfully resulted in high yields of CGTase immobilization (92.3%), activity recovery (73%) and loading efficiency (95.2%), which is undeniably one of the highest so far reported for CGTa; thus, this proposed method stands out to be a suitable candidate for industrial applications of CGTase [66].

Likewise, immobilization of nitrilase enzyme within SNPs has also been achieved by Swartz *et al.* that has shown its efficacy as a reusable biocatalyst for synthetic nitrile hydrolysis [67].

Very recently, Zakharchenko *et al.* have established a platform that coalesces/integrates two different types of core–shell-structured MNPs: one loaded with enzymes, while the other is loaded with substrate-bound therapeutic biochemicals for controlled release of chemicals or biological materials on demand [68]. This biocatalytic platform utilizes the following concepts:

- a biomimetic concept of compartmentalization;
- magnetic field-controlled transport;
- interactions of substrates and biocatalysts across semi-permeable walls of the compartments.

In this approach, the biocatalytic process is accomplished *via* magnetic field-triggered interactions when the two distinctive core–shell NPs loaded with enzyme and substrate, respectively, are brought into close/merging vicinity of each other. The enzyme and

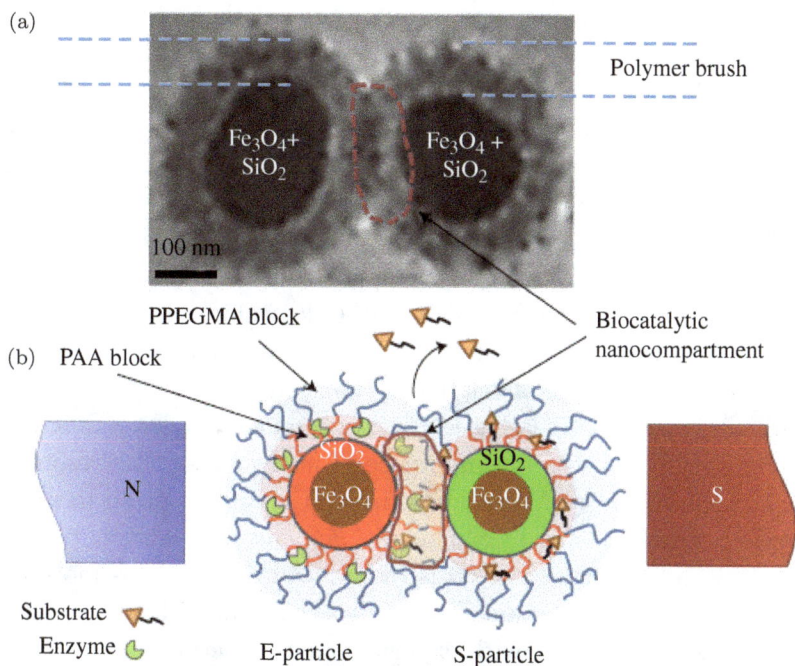

Figure 7.7. (a) A cryo-transmission electron microscopy (cryo-TEM) image of a two-particle aggregate comprising E- and S-nanoparticles and (b) schematic illustration explaining the concept of the magnetic field-triggered biocatalysis (adapted from Ref. [68]).

the substrate molecules loaded into the NPs are surrounded by polymer brushes (the semi-permeable barriers that have gating properties) that prevent their probable interaction with other competitive molecules (Figure 7.7). It is interesting to note that this reaction occurs only in the presence of a magnetic field that triggers the merging of the respective compartments.

7.3 Conclusion and Future Perspectives

In the last few decades, a plethora of well-defined silica-based nanostructured materials have been fabricated as attractive candidates for diverse applications, especially, but not limited to catalysis. Remarkable advancements have been made in the development of robust, multifunctional and recyclable nanocatalysts that impart high atom efficiency, activity, selectivity and recyclability. Despite the great successes so far, the associated numerous challenges compel to do more research in this area. The foremost challenge lies in the synthesis of identical nanostructures. Although a variety of methods are available which provide significant accuracy in controlling the size of NPs, difference in number of atoms by hundreds/thousands is observed in the individual NP. Another challenge lies in understanding the dynamics of catalytic nanostructures. NPs may undergo changes in structure, composition and electronic states during the course of reaction. Therefore, it is necessary to correlate the static and dynamic structure of active catalytic centers. Besides, structural characterization of silica-based magnetic nanocatalysts is relatively challenging due to the incompetence of nuclear magnetic resonance (NMR) spectroscopy. Moreover, emphasis should be given to understand the mechanisms at nanoscale. For this, a coalition between theorists and experimentalists is required, which involves molecular modeling, study of reaction intermediates and individual elementary catalytic steps together with experiments on reaction kinetics. Also, to bring lab-scale SNP-based catalysis at industrial scale, unification of chemists, material scientists and engineers is highly desirable. This will indeed represent a significant step toward the development of green and sustainable chemical processes.

References

[1] (a) S. B. Kalidindi, B. R. Jagirdar, *ChemSusChem* **2012**, *5*, 65–75; (b) L. L. Chng, N. Erathodiyil, J. Y. Ying, *Accounts of Chemical Research* **2013**, *46*, 1825–1837; (c) V. Polshettiwar, R. S. Varma, *Green Chemistry* **2010**, *12*, 743–754.

[2] (a) R. Sharma, S. Sharma, S. Dutta, R. Zboril, M. B. Gawande, *Green Chemistry* **2015**, *17*, 3207–3230; (b) M. B. Gawande, Y. Monga, R. Zboril, R. K. Sharma, *Coordination Chemistry Reviews* **2015**, *288*, 118–143.

[3] (a) A. Ghosh, K. Nagabhushana, D. Rautaray, R. Kumar, *Nanocatalysis Synthesis and Applications* **2013**, 643–678; (b) S. Olveira, S. P. Forster, S. Seeger, *Journal of Nanotechnology* **2014**, *2014*, 1–19.

[4] (a) M. S. Hamdy, R. Amrollahi, I. Sinev, B. Mei, G. Mul, *Journal of the American Chemical Society* **2013**, *136*, 594–597; (b) S. Wang, Z. Zhang, B. Liu, J. Li, *Catalysis Science & Technology* **2013**, *3*, 2104–2112; (c) S. Atalay, G. Ersöz, *Novel Catalysts in Advanced Oxidation of Organic Pollutants*, Springer, **2016**.

[5] (a) S. Hashiguchi, A. Fujii, J. Takehara, T. Ikariya, R. Noyori, *Journal of the American Chemical Society* **1995**, *117*, 7562–7563; (b) N. Uematsu, A. Fujii, S. Hashiguchi, T. Ikariya, R. Noyori, *Journal of the American Chemical Society* **1996**, *118*, 4916–4917.

[6] (a) J. C. Sheehan, D. Hunneman, *Journal of the American Chemical Society* **1966**, *88*, 3666–3667; (b) M. Kitamura, T. Ohkuma, S. Inoue, N. Sayo, H. Kumobayashi, S. Akutagawa, T. Ohta, H. Takaya, R. Noyori, *Journal of the American Chemical Society* **1988**, *110*, 629–631; (c) X. Cui, K. Burgess, *Chemical Reviews* **2005**, *105*, 3272–3296.

[7] (a) W. Dumont, J. C. Poulin, D. T. Phat, H. B. Kagan, *Journal of the American Chemical Society* **1973**, *95*, 8295–8299; (b) C. Bolm, A. Gerlach, *European Journal of Organic Chemistry* **1998**, *1998*, 21–27; (c) S. Kobayashi, M. Endo, S. Nagayama, *Journal of the American Chemical Society* **1999**, *121*, 11229–11230.

[8] (a) T. Ogawa, N. Kumagai, M. Shibasaki, *Angewandte Chemie International Edition* **2013**, *52*, 6196–6201; (b) Z. Chen, Z. Guan, M. Li, Q. Yang, C. Li, *Angewandte Chemie International Edition* **2011**, *50*, 4913–4917.

[9] (a) H. Hanawa, T. Hashimoto, K. Maruoka, *Journal of the American Chemical Society* **2003**, *125*, 1708–1709; (b) C. E. Song, S.-g. Lee, *Chemical Reviews* **2002**, *102*, 3495–3524.

[10] R. Dalpozzo, *Green Chemistry* **2015**, *17*, 3671–3686.

[11] (a) J. M. Thomas, R. Raja, *Accounts of Chemical Research* **2008**, *41*, 708–720; (b) M. Heitbaum, F. Glorius, I. Escher, *Angewandte Chemie International Edition* **2006**, *45*, 4732–4762; (c) C. Li, H. Zhang, D. Jiang, Q. Yang, *Chemical Communications* **2007**, 547–558; (d) M. Bartók, *Chemical Reviews* **2009**, *110*, 1663–1705.

[12] H. Zhang, R. Jin, H. Yao, S. Tang, J. Zhuang, G. Liu, H. Li, *Chemical Communications* **2012**, *48*, 7874–7876.

[13] H. L. Ngo, W. Lin, *The Journal of Organic Chemistry* **2005**, *70*, 1177–1187.

[14] D. J. Mihalcik, W. Lin, *Angewandte Chemie International Edition* **2008**, *47*, 6229–6232.

[15] J. Li, Y. Zhang, D. Han, Q. Gao, C. Li, *Journal of Molecular Catalysis A: Chemical* **2009**, *298*, 31–35.

[16] (a) F. Jamali, R. Mehvar, F. Pasutto, *Journal of Pharmaceutical Sciences* **1989**, *78*, 695–715; (b) J. M. Williams, R. J. Parker, C. Neri, *Enzyme Catalysis in Organic Synthesis: A Comprehensive Handbook, Second Edition* **2002**, 287–312.

[17] H. M. Gardimalla, D. Mandal, P. D. Stevens, M. Yen, Y. Gao, *Chemical Communications* **2005**, 4432–4434.

[18] S. Luo, X. Zheng, J.-P. Cheng, *Chemical communications* **2008**, 5719–5721.

[19] H. C. Kolb, M. S. VanNieuwenhze, K. B. Sharpless, *Chemical Reviews* **1994**, *94*, 2483–2547.

[20] D. Lee, J. Lee, H. Lee, S. Jin, T. Hyeon, B. M. Kim, *Advanced Synthesis & Catalysis* **2006**, *348*, 41–46.

[21] A. Schätz, M. Hager, O. Reiser, *Advanced Functional Materials* **2009**, *19*, 2109–2115.

256   *Silica-Based Organic–Inorganic Hybrid Nanomaterials*

[22] J. Lim, S. N. Riduan, S. S. Lee, J. Y. Ying, *Advanced Synthesis & Catalysis* **2008**, *350*, 1295–1308.
[23] C. Grondal, M. Jeanty, D. Enders, *Nature Chemistry* **2010**, *2*, 167–178.
[24] (a) K. Nicolaou, D. J. Edmonds, P. G. Bulger, *Angewandte Chemie International Edition* **2006**, *45*, 7134–7186; (b) K. Nicolaou, J. S. Chen, *Chemical Society Reviews* **2009**, *38*, 2993–3009.
[25] Y. Huang, S. Xu, V. S. Y. Lin, *Angewandte Chemie International Edition* **2011**, *50*, 661–664.
[26] P. Li, Y. Yu, H. Liu, C.-Y. Cao, W.-G. Song, *Nanoscale* **2014**, *6*, 442–448.
[27] S. Dutta, S. Sharma, A. Sharma, R. K. Sharma, *ACS Omega* **2017**, *2*, 2778–2791.
[28] E. Forgacs, T. Cserhati, G. Oros, *Environment International* **2004**, *30*, 953–971.
[29] T. Robinson, G. McMullan, R. Marchant, P. Nigam, *Bioresource Technology* **2001**, *77*, 247–255.
[30] U. Akpan, B. Hameed, *Journal of Hazardous Materials* **2009**, *170*, 520–529.
[31] F. Wang, M. Li, L. Yu, F. Sun, Z. Wang, L. Zhang, H. Zeng, X. Xu, *Scientific Reports* **2017**, *7*, 6960–6969.
[32] A. Ajmal, I. Majeed, R. N. Malik, H. Idriss, M. A. Nadeem, *RSC Advances* **2014**, *4*, 37003–37026.
[33] R. Singh, R. Bapat, L. Qin, H. Feng, V. Polshettiwar, *ACS Catalysis* **2016**, *6*, 2770–2784.
[34] S.-K. Li, F.-Z. Huang, Y. Wang, Y.-H. Shen, L.-G. Qiu, A.-J. Xie, S.-J. Xu, *Journal of Materials Chemistry* **2011**, *21*, 7459–7466.
[35] D. Greene, R. Serrano-Garcia, J. Govan, Y. K. Gun'ko, *Nanomaterials* **2014**, *4*, 331–343.
[36] J. Cheng, R. Ma, M. Li, J. Wu, F. Liu, X. Zhang, *Chemical Engineering Journal* **2012**, *210*, 80–86.
[37] S. Sarina, E. R. Waclawik, H. Zhu, *Green Chemistry* **2013**, *15*, 1814–1833.
[38] (a) X. Liang, P. Wang, M. Li, Q. Zhang, Z. Wang, Y. Dai, X. Zhang, Y. Liu, M.-H. Whangbo, B. Huang, *Applied Catalysis B: Environmental* **2018**, *220*, 356–361; (b) S. Glaus, G. Calzaferri, R. Hoffmann, *Chemistry — A European Journal* **2002**, *8*, 1785–1794.
[39] K. D. Rasamani, J. J. Foley, Y. Sun, *Nano Futures* **2018**, *2*, 015003– 015015.
[40] (a) B. Liu, Z. Zhang, *ACS Catalysis* **2015**, *6*, 326–338; (b) D. L. Klass, *Biomass for Renewable Energy, Fuels, and Chemicals*, Elsevier, **1998**; (c) M. Demirbas, M. Balat, *Energy Conversion and Management* **2006**, *47*, 2371–2381; (d) S. N. Naik, V. V. Goud, P. K. Rout, A. K. Dalai, *Renewable and Sustainable Energy Reviews* **2010**, *14*, 578–597; (e) P. M. Schenk, S. R. Thomas-Hall, E. Stephens, U. C. Marx, J. H. Mussgnug, C. Posten, O. Kruse, B. Hankamer, *Bioenergy Research* **2008**, *1*, 20–43.
[41] D.-M. Lai, L. Deng, Q.-X. Guo, Y. Fu, *Energy & Environmental Science* **2011**, *4*, 3552–3557.
[42] (a) Y. H. P. Zhang, L. R. Lynd, *Biotechnology and Bioengineering* **2004**, *88*, 797–824; (b) G. W. Huber, S. Iborra, A. Corma, *Chemical Reviews* **2006**, *106*, 4044–4098.

[43] Y. Xiong, Z. Zhang, X. Wang, B. Liu, J. Lin, *Chemical Engineering Journal* **2014**, *235*, 349–355.

[44] (a) S. Xiao, B. Liu, Y. Wang, Z. Fang, Z. Zhang, *Bioresource Technology* **2014**, *151*, 361–366; (b) R.-J. van Putten, J. C. van der Waal, E. De Jong, C. B. Rasrendra, H. J. Heeres, J. G. de Vries, *Chemical Reviews* **2013**, *113*, 1499–1597; (c) T. Pasini, M. Piccinini, M. Blosi, R. Bonelli, S. Albonetti, N. Dimitratos, J. A. Lopez-Sanchez, M. Sankar, Q. He, C. J. Kiely, *Green Chemistry* **2011**, *13*, 2091–2099; (d) B. Saha, D. Gupta, M. M. Abu-Omar, A. Modak, A. Bhaumik, *Journal of Catalysis* **2013**, *299*, 316–320.

[45] (a) A. Gandini, N. M. Belgacem, *Polymer International* **1998**, *47*, 267–276; (b) M. Del Poeta, W. A. Schell, C. C. Dykstra, S. Jones, R. R. Tidwell, A. Czarny, M. Bajic, M. Bajic, A. Kumar, D. Boykin, *Antimicrobial Agents and Chemotherapy* **1998**, *42*, 2495–2502; (c) A. S. Amarasekara, D. Green, L. D. Williams, *European Polymer Journal* **2009**, *45*, 595–598; (d) K. T. Hopkins, W. D. Wilson, B. C. Bender, D. R. McCurdy, J. E. Hall, R. R. Tidwell, A. Kumar, M. Bajic, D. W. Boykin, *Journal of Medicinal Chemistry* **1998**, *41*, 3872–3878.

[46] S. Wang, Z. Zhang, B. Liu, J. Li, *Industrial & Engineering Chemistry Research* **2014**, *53*, 5820–5827.

[47] I. Podolean, V. Kuncser, N. Gheorghe, D. Macovei, V. I. Parvulescu, S. M. Coman, *Green Chemistry* **2013**, *15*, 3077–3082.

[48] (a) J. Zeikus, M. Jain, P. Elankovan, *Applied Microbiology and Biotechnology* **1999**, *51*, 545–552; (b) I. Bechthold, K. Bretz, S. Kabasci, R. Kopitzky, A. Springer, *Chemical Engineering & Technology* **2008**, *31*, 647–654.

[49] A. Cukalovic, C. V. Stevens, *Biofuels, Bioproducts and Biorefining* **2008**, *2*, 505–529.

[50] K. B. Sidhpuria, A. L. Daniel-da-Silva, T. Trindade, J. A. Coutinho, *Green Chemistry* **2011**, *13*, 340–349.

[51] Y.-Y. Lee, K. C.-W. Wu, *Physical Chemistry Chemical Physics* **2012**, *14*, 13914–13917.

[52] W.-H. Peng, Y.-Y. Lee, C. Wu, K. C.-W. Wu, *Journal of Materials Chemistry* **2012**, *22*, 23181–23185.

[53] (a) M. T. Reetz, *Journal of the American Chemical Society* **2013**, *135*, 12480–12496; (b) A. Liese, L. Hilterhaus, G. Antranikian, U. Kettling, *Applied Biocatalysis: From Fundamental Science to Industrial Applications*, John Wiley & Sons, **2016**.

[54] A. Schmid, J. Dordick, B. Hauer, A. Kiener, M. Wubbolts, B. Witholt, *Nature* **2001**, *409*, 258.

[55] (a) D. J. Pollard, J. M. Woodley, *Trends in Biotechnology* **2007**, *25*, 66–73; (b) J. A. Tao, G.-Q. Lin, A. Liese, *Biocatalysis for the Pharmaceutical Industry: Discovery, Development, and Manufacturing*, John Wiley & Sons, **2009**.

[56] (a) M. T. Reetz, *Advanced Materials* **1997**, *9*, 943–954; (b) R. A. Sheldon, S. van Pelt, *Chemical Society Reviews* **2013**, *42*, 6223–6235; (c) J. Kim, J. W. Grate, P. Wang, *Chemical Engineering Science* **2006**, *61*, 1017–1026.

[57] (a) S. R. Patel, M. G. Yap, D. I. Wang, *Biochemical Engineering Journal* **2009**, *48*, 13–21; (b) Y.-X. Bai, Y.-F. Li, Y. Yang, L.-X. Yi, *Process Biochemistry* **2006**, *41*, 770–777; (c) M. I. Kim, H. O. Ham, S.-D. Oh, H. G. Park, H. N. Chang, S.-H. Choi, *Journal of Molecular Catalysis B: Enzymatic* **2006**, *39*, 62–68; (d) M. Kalantari, M. Yu, Y. Yang, E. Strounina, Z. Gu, X. Huang, J. Zhang, H. Song, C. Yu, *Nano Research* **2017**, *10*, 605–617; (e) M. C. Silva, J. A. Torres, F. G. Nogueira, T. S. Tavares, A. D. Corrêa, L. C. Oliveira, T. C. Ramalho, *RSC Advances* **2016**, *6*, 83856–83863.

[58] M. T. Reetz, A. Zonta, V. Vijayakrishnan, K. Schimossek, *Journal of Molecular Catalysis A: Chemical* **1998**, *134*, 251–258.

[59] J. Lee, Y. Lee, J. K. Youn, H. B. Na, T. Yu, H. Kim, S. M. Lee, Y. M. Koo, J. H. Kwak, H. G. Park, *Small* **2008**, *4*, 143–152.

[60] H.-H. Yang, S.-Q. Zhang, X.-L. Chen, Z.-X. Zhuang, J.-G. Xu, X.-R. Wang, *Analytical Chemistry* **2004**, *76*, 1316–1321.

[61] Y. Kuwahara, T. Yamanishi, T. Kamegawa, K. Mori, M. Che, H. Yamashita, *Chemical Communications* **2012**, *48*, 2882–2884.

[62] Y. C. Lee, C. T. Chen, Y. T. Chiu, K. C. W. Wu, *ChemCatChem* **2013**, *5*, 2153–2157.

[63] V. Califano, F. Sannino, A. Costantini, J. Avossa, S. Cimino, A. Aronne, *The Journal of Physical Chemistry C* **2018**, *122*, 8373–8379.

[64] S. Sommaruga, E. Galbiati, J. Peñaranda-Avila, C. Brambilla, P. Tortora, M. Colombo, D. Prosperi, *BMC Biotechnology* **2014**, *14*, 82.

[65] J. C. Cruz, K. Würges, M. Kramer, P. H. Pfromm, M. E. Rezac, P. Czermak, *Nanoscale Biocatalysis*, Springer, **2011**, pp. 147–160.

[66] A. S. Ibrahim, A. A. Al-Salamah, A. M. El-Toni, M. A. El-Tayeb, Y. B. Elbadawi, *Electronic Journal of Biotechnology* **2013**, *16*, 1–16.

[67] J. D. Swartz, S. A. Miller, D. Wright, *Organic Process Research & Development* **2009**, *13*, 584–589.

[68] A. Zakharchenko, N. Guz, A. M. Laradji, E. Katz, S. Minko, *Nature Catalysis* **2018**, *1*, 73–81.

Index

www.ingramcontent.com/pod-product-compliance
Lightning Source LLC
Chambersburg PA
CBHW050549190326
41458CB00007B/1975